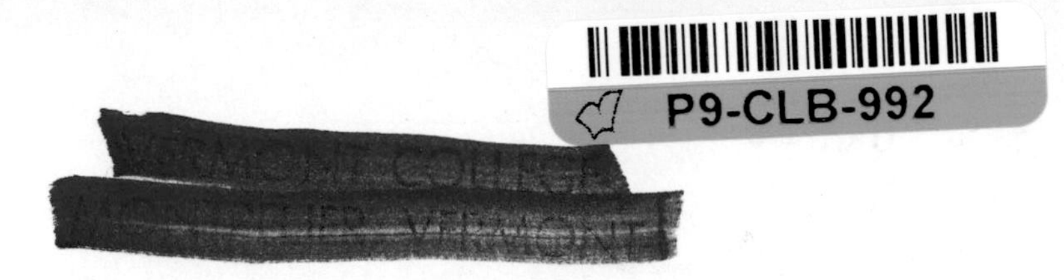

# SOCIAL BEHAVIOR AND ORGANIZATION AMONG VERTEBRATES

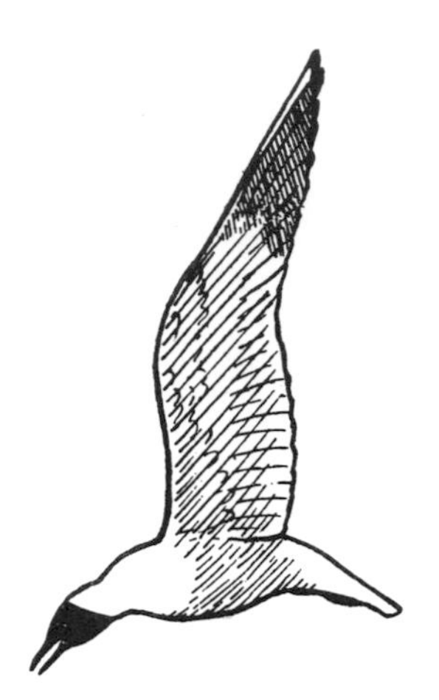

THE UNIVERSITY OF CHICAGO PRESS / Chicago and London

# SOCIAL BEHAVIOR AND ORGANIZATION AMONG VERTEBRATES



*Edited by*

WILLIAM ETKIN

Library of Congress Catalog Card Number: 64-13947

THE UNIVERSITY OF CHICAGO PRESS, CHICAGO AND LONDON
The University of Toronto Press, Toronto 5, Canada

Published 1964. Third Impression 1967
Printed in the United States of America

*Contributors:*

FRANK A. BEACH

DAVID E. DAVIS

WILLIAM ETKIN

DANIEL S. LEHRMAN

J. P. SCOTT

NIKO TINBERGEN

# Preface

In the last quarter century there has been a burgeoning of interest among zoologists in the study of social behavior in animals. This interest has been spreading to experimental and clinical psychologists and to other students of human behavioral sciences as the relevance of the zoological findings to the other sciences becomes apparent. The increased concern with animal social behavior is shown by the outpouring of books and articles intended for popular consumption and by the inclusion of studies in animal social behavior in the programs of professional societies in many areas. Yet no introduction to the field intended for serious students of the zoological or human behavioral sciences is presently available. This book is intended to fill that void. Limitations of space made it necessary to restrict the discussion to the vertebrate group.

The present work is organized around two approaches. An analytical description of social behavior from the ecological and evolutionary point of view is given in a series of four chapters written by the editor. As these chapters are presented from a non-technical viewpoint and assume only an elementary acquaintance with biological concepts they should serve to introduce the non-zoologist to this field. The second approach is experimental. Five chapters are contributed by active experimentalists; each author covers recent developments in the area of his own specialty. Practical considerations made it necessary to limit the number of participants invited to this section, but it was the purpose of the editor to select representatives to cover the major areas of current research in English-speaking laboratories. It is hoped that the interested reader will be informed, not only as to the status of contemporary analysis, but also specifically as to the work of the many investigators now active in this field. Though these chapters should appeal most directly to zoology, psychology, and psychiatry students, it is hoped that they will also inspire students interested in the social sciences to perceive the relevance and interest of the experimental work now being done in the animal field. An introductory chapter summarizing recent developments in endocrinology and neurology in so far as these are directly relevant to the experimental analysis has been included in order to bring the student up to date in this area.

The concept of this book was developed with the textbook committee of the Section of Animal Behavior of the American Society of Zoologists beginning in 1956. The editor, as chairman of the committee, is indebted to numerous members of this section for valuable discussions of the project, in particular to J. P. Scott and A. M. Guhl, chairmen of the section at the time, and to Martin Schein, John Emlen, Edgar Hale, Fritz Hilton, John A. King, Edward Wilson, and others in and out of the committee. The actualization of the book, however, had to wait until the active support of the Interdisciplinary Program related to the Nervous System and Mental Functions at the Albert Einstein College of Medicine made it possible to bring the authors together for seminars at that institution. For the vigorous and far-sighted support in this effort, the editor wishes to express his feeling of sincere gratitude to Ernst Scharrer and to the late Saul Korey. Thanks are also due to the Ford Foundation which generously provided a Faculty Fellowship in 1954, which enabled the editor to develop a course in social behavior of animals which he has been giving at the City College since then. The many friends and colleagues whose encouragement has urged the editor on must be thanked anonymously, but the assistance and forbearance of his wife, so indispensable to the long task, require an individual vote of thanks.

WILLIAM ETKIN

New York

# Contents

# Illustrations

WILLIAM ETKIN

# 1

# Co-operation and Competition in Social Behavior

## Natural Selection

### DARWINIAN CONCEPTS

The triumph of the theory of evolution and its Darwinian explanation in the late nineteenth century had repercussions in almost every field of thought. Not the least of these followed the application of the Darwinian ideas of competition and struggle for existence to the social life of man. Economists, sociologists, and social philosophers sought to explain social forms and practices in terms of their contributions to the success and survival of the group. In academic circles, such thinking led to the theories of the Social Darwinists. One of the main conclusions of these theorists was that it is necessary to permit the strong to exploit the weak, for by this action human progress is produced. As we look back now, we may perhaps view these theories as attempts to justify the harsh attitudes of laissez faire capitalism toward the social evils of the day (Hofstadter, 1944).

Whereas we will not be directly concerned with the development of sociological theory in this book, we must take notice of these ideas, because both their direct effect and particularly the reaction against them strongly colored contemporary thought on social behavior in animals. On the one hand, the significance, and even the relevance, of animal studies to an understanding of man has been denied in the

William Etkin is professor of biology at the City College of New York and visiting research professor of anatomy at the Albert Einstein College of Medicine, New York.

effort to reject the fancied consequences for man. On the other hand, great efforts have been made to counteract the nineteenth-century interpretation of Darwinism at the animal level by denigrating the role of natural selection. We shall see that both viewpoints suffer from an inadequate appreciation of the complexity of animal social life. For a proper appreciation of the evolutionary aspects of our subject, however, it would be well to begin with a brief consideration of some of the biological fundamentals involved in our present understanding of natural selection.

The concept of natural selection, popularly called survival of the fittest, has had a stormy history since Darwin's day but has emerged clarified and greatly strengthened in contemporary biology. It may be briefly characterized as follows. The reproductive capacity of any species is so high that the increase in population tends to outrun the available necessities such as food and shelter. In the resulting competition some members of the species, being better endowed by hereditary characteristics, survive and reproduce more than the others. Because of their success the hereditary factors or genes which they carry tend to be preserved and passed on in greater and greater measure to succeeding generations. Through this process, the pool of genes characteristic of the species tends to shift in the direction of greater adaptability. A species long established in a particular environment reaches an optimal balance among its genetic factors, but with each alteration in the environment the balance of these factors in the population tends to shift. It is important to note that the competition upon which natural selection acts is chiefly that between members of the same species, because members of different species, having different modes of making a living and occupying different ecological niches, do not compete as directly with one another.

In the latter part of the nineteenth century many biologists doubted that natural selection was adequate to explain the facts of evolution. In fact Darwin himself was not exclusively committed to natural selection as the driving power of evolution. In particular he often fell back upon the inheritance of acquired characteristics for an explanation of specific instances of evolutionary change. This concept, sometimes referred to as Lamarckianism, after Lamarck, the eighteenth-century French zoologist who espoused it, maintains that modifications in the body tissues which are acquired during an animal's lifetime induce tendencies for similar modifications to be transmitted to its offspring. In relation to behavior, the idea of inheritance of acquired characteristics suggests that habits learned by experience in one generation tend to become "built into" the germ plasm as "instincts" in subsequent evolution.

Contemporary biology, however, has discarded the Lamarckian concept. For one thing, experimental tests of its claims have been overwhelmingly negative. For another, the development of modern genetics has given us clear insights into the mechanisms of inheritance, and, as these are now understood, they do not provide any way in which environmentally induced modifications in the body could influence the genes. Finally, biologists today find no need to fall back upon the Lamarckian hypothesis to explain evolution. According to our present understanding of the genetics of animal populations, the gene pool of a population is a balanced system wonderfully sensitive to selection pressures. The double set of genes in each individual (diploid condition) allows the accumulation of gene mutations in the population. These furnish the raw material upon which natural selection acts. Sexual reproduction permits rapid diffusion of genetic change throughout a population. Together these factors make natural selection speedier and more efficient than anything previously suspected by biologists, even early in the twentieth century. As a consequence, the old objection that natural selection is too slow and ineffective to account for evolutionary change has lost its cogency. Experimental tests have supported the modern view that natural selection operating on the genetic system of higher organisms can effect rapid and delicate adaptation of these organisms to the changing demands of the environment (Dobzhansky, 1951).

The above viewpoint is not controverted by the finding that other factors, such as group isolation and random variations in small populations, are also of some importance in determining evolutionary change and may, on occasion, lead to developments not fundamentally adaptive in nature. These factors operate as modifying influences to the driving force of natural selection and, whatever their importance in minor evolutionary change, do not displace the long-term adaptive nature of the evolutionary process as produced by the action of natural selection. In conclusion we may therefore say that the contemporary view of evolution is that it is guided and directed by natural selection. In the large view the characteristics of organisms must be expected to be adaptive in the sense of contributing to the long-run reproductive efficacy of the species as it lives in its own particular ecological niche.

## SOCIALITY AND DARWINISM

Before we examine the implications of natural selection for social behavior, it would be well to clarify some terms. By "social response" we shall mean a behavioral response made by a group-living animal to another in its group but not to animals or objects which are not members of the group. We shall regard groups as "social" when the members stay together as a result of their social responses to one another rather than by responses to other factors in their environment. Groups that are held together by responses to such other factors will be called "aggregations." Thus a flock of sheep is a social group, since it is maintained by the social responses of the animals to one another; but the massing of insects around a light at night is an aggregation, since it results from their common attraction to the light. Of course, in many cases we are unable to decide, because of lack of appropriate evidence, whether a group is truly social or not, and the noncommittal term "group" may then be used.

Granted that modern biology accepts the primary importance of natural selection in evolution, the question remains: "What is it that has survival value?" It is obvious that what may be called aggressive potential, that is, the ability to win in a conflict with competitors whether by wit, bluff, or brute force, is often the deciding factor in survival. Yet if survival were possible only for the most aggressive and self-seeking individuals, it is difficult to see how social life could develop far among animals. But despite this difficulty social life is evidently very common among higher animals, particularly vertebrates. This is impressed upon us if we merely think for a moment of the large number of words that our language provides to describe the naturally occurring groups of higher animals. Thus of birds we say a flock of robins, a gaggle of geese, a covey of quail, a bevy of pheasants, a brood of ducklings, a hatch of chicks, a set of swans. Of mammals we say a flock of sheep, a herd of elk, a pride of lions, a litter of puppies, a pack of wolves, a farrow of pigs, a colony of monkeys or gophers, a school of whales.

In a book on social life in animals in 1927, Alverdes summarized the knowledge at that time of the occurrence of group life among animals. He classified and described at some length the kinds of social organization found among animals. From this work we gather an impressive view of the varied and widespread character of animal sociality. Indeed, if we take a literal view of what constitutes social behavior as defined above (i.e., distinctive responses between members of a group which tend to keep them together) and therefore include in it the interaction of the sexes in reproduction, we may regard social behavior as universal among vertebrates, since all reproduce sexually. But aside from the special question of sexual behavior, it is still evident that there is much more social life to be found in the animal world than the theory of natural selection, with its emphasis on competition, might suggest.

It is clear that the well-established social group affords many advantages to its members. When a good food source is found by one member, all are attracted to it by the behavior of the finder. Animals in groups are much less subject to predation than are animals in isolation. The individuals warn one another of approaching danger so that it is very difficult for a predator to get close to a group (Fig. 1.1). Group action against a predator, as we find it in the "mobbing" of a hawk by song birds and in defensive formations in ungulates, provides protection for the members. The musk oxen form a defensive

FIG. 1.1. When the pronghorn antelope detects danger, the hairs of the white rump patch are erected, producing a conspicuous white patch said to be visible for miles on the open plains. When one such animal dashes away, the flashing white signal alerts the others even if the herd is widely scattered. Such protective devices greatly increase the security of the herd. (After E. T. Seton, *Lives of Game Animals,* Vol. 3; by permission of Charles T. Branford Co., Publishers.)

ring with large males on the outside. Group formation, as in close flocking in birds is so effective in protecting the individual that, generally speaking, predators try to isolate an animal from a group before attacking it (Fig. 1.2). Group life also facilitates reproduction in many ways. Some of these, such as the bringing together of the sexes at the appropriate time, are obvious, but some which we will discuss later are rather subtle.

Undoubtedly, therefore, genes which make for co-operative behavior convey some survival value to their possessors by encouraging group living. Many naturalists and philosophers in the post-Darwinian era emphasized this fact. They tried to show that not only the competitive survive, but that co-operative behavior, too, is found among

FIG. 1.2. A flock of starlings before a falcon appeared (*left*) and after (*right*) is shown here. The tendency of group-living animals to close ranks when danger appears is widespread throughout the animal kingdom. When the prey is in close order the falcon cannot attack them, since it would be in danger of injuring itself. (After N. Tinbergen, *Study of Instinct*, 1949.)

animals and has survival value. The Russian naturalist Kropotkin wrote the most notable of the books along these lines, *Mutual Aid* (1914). In it he collected instances, often anecdotal, of acts of co-operation and of individual self-sacrifice among animals and argued strongly that co-operative behavior is an important factor in evolution.

In spite of the earnest and well-meant efforts to show that co-operative behavior has survival value, the main force of natural selection must be expected to favor self-seeking, "anti-social" actions by the individual. This clearly follows from the point made above that natural selection operates chiefly in the competition between members of the same species and particularly between individuals closely associated with one another. In such competition the survivors are inevitably those individuals which gratify their needs at the expense of other members of the group or species. It was difficult for theoretic evolutionists to see how co-operative behavior could develop very far in the face of this situation.

A factor introducing some confusion into our thinking in this area arises in the special case of social insects which are well known to show self-sacrificing behavior on the part of individuals. The worker bee that stings an intruder dies as a result of her action taken in defense of the hive, since her entrails are pulled out with the sting. However, in insect groups such as ants and bees where reproductive activity is confined to one pair in the colony, the queen and her mate, the action of natural selection is very different from what it is in vertebrates. In these insects the co-operative behavior of the workers not only facilitates group survival but also insures transmission of the genes that promote this behavior. Such behavior preserves the queen, who carries genes making for co-operative behavior in her germ cells and who passes these genes on, not only to her workers, but to her queen daughters as well. Thus the bee that sacrifices itself in stinging an enemy of the hive actually helps insure the transmission of the genes which favor such behavior.

But this very example indicates the weakness of the position of those who argue from the behavior of social insects to that of social vertebrates. For, in vertebrates, reproduction is a function of all members of the group. In contrast to the situation in insects, an individual which sacrifices itself in protection of the group insures that, in general, its own genes will *not* be passed on to the next generation. On the whole then, one would expect from such reasoning that the evolutionary pressure for "selfish" behavior would be considerably stronger than that for co-operation and that co-operative behavior would be much less conspicuous than its opposite in vertebrates.

But whichever of these views one might favor on theoretic grounds, it is clear that this type of speculative analysis leads to no definite conclusion. It is difficult to reconcile the action of natural selection with the widespread occurrence of social organization in vertebrates. The explanation for group formation offered by Alverdes was that social animals have a social instinct absent in non-social creatures. We shall see that this idea of a social instinct hides a multiplicity of interacting factors, many of them dependent upon experience rather than being innate as implied by the term instinct. It will be our task in this book to analyze the many factors which modern study of animal behavior has uncovered, especially in the last quarter of a century or so. We shall see that older ideas of the action of natural selection were hopelessly naïve and hid much of the subtlety and variety of animal life under vague generalization. For example, later in this chapter we shall see how aggressiveness itself has been so modified by social hierarchy and territorialism as to reinforce rather than disrupt social life. In the light of modern studies, the social life of animals is seen to be vastly more complex and intricately balanced than envisaged in the discussions of post-Darwinian naturalists. Before we examine hierarchy and territory, however, we must consider an aspect of co-operative behavior brought out in the experimental work of recent years.

## Physiological Effects of Aggregations

In the second quarter of the present century, an experimental search for the basis of animal sociality was vigorously pursued by W. C. Allee and his students (Allee, 1951). They investigated the physiological effects of grouping among various non-social animals. This approach was suggested by the fact that such animals are often found in temporary aggregations in nature. These workers uncovered a number of favorable "group effects." Some of these involve a chemical or physical change in the environment produced by the members of the group. For example, if aquatic animals, such as goldfish or flatworms, are grown in unfavorable natural waters or water into which small quantities of harmful chemicals have been introduced, they do better when combined into small groups than as isolated individuals. In particular instances the improvement has been found to result from the neutralization of poisons by the body secretions of the animals. In other cases the loss of salts to the water by animals living in it was the basis of the beneficial effects observed. Water which has been improved by organisms living in it is said to be "conditioned," and its use in fish culture by aquarists has become a common practice. In other instances, as, for example, in mice raised in a cold environment, the advantage of group living has been traced to the heat and shelter provided by the animals' own bodies. In still other examples the beneficial group effect has been found to be based on behavioral or psychological characteristics, as in the case of the reduction in metabolism in snakes that are allowed to aggregate as compared to the same animals kept apart (Fig. 1.3). Members of a group were also found to stimulate one another to greater activity because of a tend-

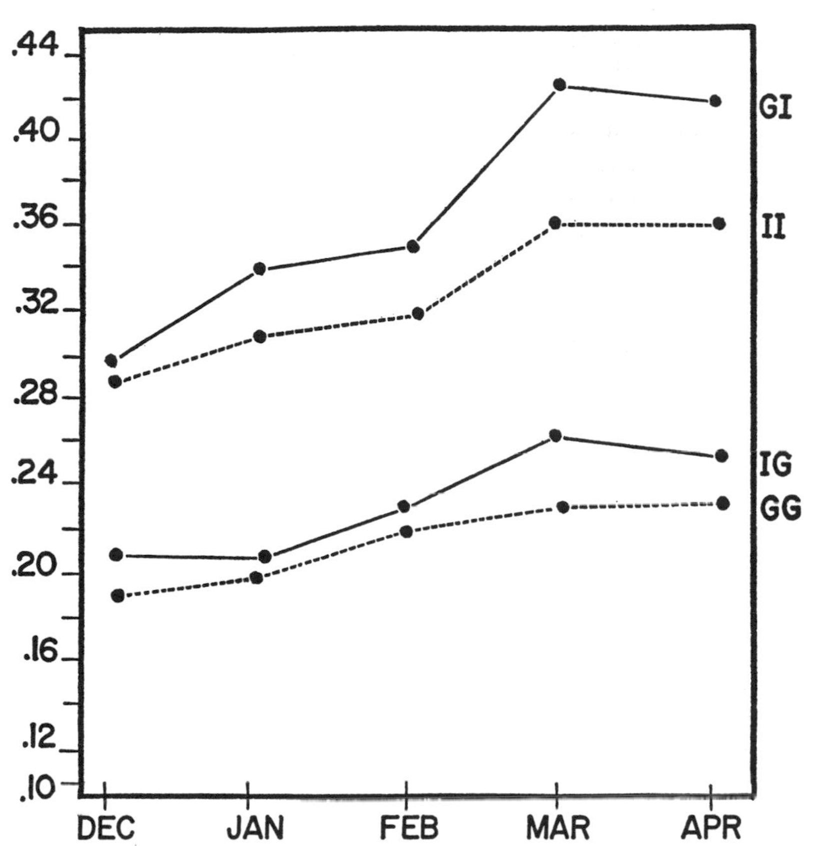


FIG. 1.3. The effects of aggregation and isolation on the rate of oxygen consumption of brown deKay snakes is shown here. One set of animals was kept in a group and tested in the grouped condition (GG) and also tested in isolation (G*I*). Another set was kept in isolation and tested grouped (*I*G) and again in isolation (*II*). The results show that animals tested while grouped (*I*G and GG) show lower oxygen consumption than the isolated animals, irrespective of the previous condition of the animals. Ordinates represent cubic centimeters oxygen per gram weight per hour. Abscissae represent monthly averages. (After J. Clausen, *Cell. and Comp. Physiol.,* 8 [1936].)

ency to imitate (social facilitation). Experiments demonstrate such social facilitation in respect to eating, resting, and learning activity in animals as diverse as sunfish (Fig. 1.4) and children (Allee, 1951).

Perhaps the most biologically significant finding in these studies is the evidence that isolated cells give off substances into the medium in which they are grown which act as stimulants to the growth of other cells. This phenomenon has not as yet been adequately clarified, and the reality and nature of the hypothetical substances is still in

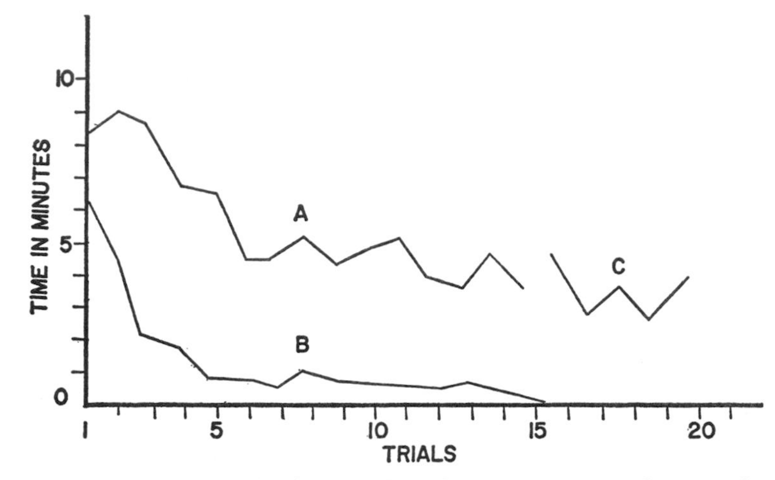


FIG. 1.4. The social facilitation of performance in a simple maze by green sunfish is shown in the graph above. *Curve A* shows the average time required to pass through the apparatus by thirty isolated normal fish. *Curve B* indicates the time required for thirty animals grouped three at a time. After the fifteenth trial the grouped animals were tested as isolates (*Curve C*). Since these animals showed the same performances as those continuously isolated, it is clear that their previous grouping had not resulted in greater learning. The facilitation must therefore have resulted from the leadership furnished by the fastest animals in the groups. (After E. B. Hale, *Physiol. Zool.* 29 [1956].)

doubt. If found to be of general occurrence, this phenomenon might be basic in the physiology and evolution of multicellular organisms.

From this work we can see that there are sometimes subtle selective benefits that accrue to animals by virtue of being aggregated into groups. We can appreciate, therefore, that there are evolutionary pressures, previously unsuspected, for the formation of social groups among non-social animals. Yet though such factors may have played a part in the origin of social life among animals, it is doubtful that they are strong enough or sufficiently general to be of great significance in explaining the existence of the more complex social organizations. In any case, though this work shows us something of the complexity of the factors involved in the survival value of group life, it tells us nothing of the physiological mechanisms which operate in maintaining the integrity of social groups.

As an example of how little we understand of the physiological aspect of group life in the less highly developed societies, we may briefly consider schooling behavior in fish. This is a very widespread phenomenon among fish, particularly such as live in the open sea—

herring, for example. Schools of fish generally consist of animals of similar size, often all of one species. Yet we know little about the benefits this schooling habit affords the animals. Many suggestions have been made: that the animals are protected from predators by the confusing effect of their great numbers, that they economize in energy required for swimming by keeping in the slip-stream of the nearby fish, etc. But these have not been experimentally demonstrated. We do know from experiment, however, that some schooling fish are visually attracted to any moving object that is neither too large nor moving too rapidly but are repelled by the same objects at close quarters. These responses probably account for the regular spacing of the fish

FIG. 1.5. Jacks schooling in large marine tank. Note the uniformity of spacing maintained by the fish. (Drawing adapted from a photograph, courtesy of Marine studios.)

in the school (Fig. 1.5). But the pattern and mechanisms of these responses have not yet been extensively analyzed because of the obvious difficulty of experimenting on animals that normally live in the open sea (Shaw, 1962). The role of social facilitation and other factors in the flocking of birds was emphasized by Crook (1961).

## Individuality and Dominance

### INDIVIDUAL RECOGNITION

Schjelderup-Ebbe, in his studies of bird flocks (see his summary, 1935), made the fundamental observation that the members of small flocks of many species of birds recognize one another as individuals. This observation was the outgrowth of the fact that his study of their group life was intensive enough so that he himself was able to recognize the individual members. This basic methodological insight—that the experimenter must, by means such as artificial marking and others, identify the individual in a group in order to understand its behavior —has become the basis of a host of studies that generally confirmed and extended Schjelderup-Ebbe's conclusion, which we may quote in his own colorful language:

> Every bird is a personality. However, the word "personality" must not be understood to have the same meaning in this statement as when used in regard to human beings. What is meant here is that any one bird, irrespective of the species to which it belongs, is absolutely distinct in character and in the manifestations of character from any other bird of its species.
>
> This may sound odd, but that is only because the individual and social psychology of birds has been regarded too superficially. No attempt has been made to know each individual bird in a given flock. So to know them, however, is the most important prerequisite for the full understanding of the general and comparative psychology of birds. The ability to distinguish each individual provides the key for the solution of a series of problems which we should otherwise be unable to solve and which are not only of ornithological interest but also of importance for the understanding of the general continuity which prevails in all life.

This characteristic of individuality in animals, upon which, as seen above, Schjelderup-Ebbe placed the greatest emphasis, is strikingly manifested in many aspects of group behavior. For example, in a huge colony of gannets, gulls, or other sea birds, each individual recognizes its own mate and behaves quite differently to it as it approaches the nest than it does to other members of the flock. To the human observer the birds may look and sound alike, yet such a bird shows recognition of its mate when it is fifty or more feet away. In parent-offspring relations, individual recognition is often strikingly shown (although in the commonly studied laboratory animals, the rat and mouse, it is markedly absent). In most herd animals the mother-child relation is highly specific; in the fur seal, for example, among the welter of young moving about on the beach, a mother returning from a feeding expedition of several days unerringly picks her own child and will give suck to no other (Bartholomew, 1953).

Schjelderup-Ebbe's studies, however, were largely confined to the competitive aspects of individual recognition in birds. In this connection, individual recognition gives rise to the phenomenon of behavioral dominance or, to use Schjelderup-Ebbe's term, "despotism." It is perhaps unfortunate that the prominence given to dominance behavior in recent studies has tended to obscure the more general significance of Schjelderup-Ebbe's contribution in respect to individuality. Though we shall have repeated occasion to refer to the individuality of members of an animal group in other connections, we are here concerned with it, as Schjelderup-Ebbe was, in connection with intra-group competition. Studies along these lines have been carried out most extensively with birds, particularly with domestic chickens. Our initial comments will be based, therefore, largely on these studies.

### HIERARCHY

In a small group of hens (up to about ten members) which have been living together with ample space for some time, very little aggressive behavior is ordinarily seen. Chickens are commonly fed by scattering the grain widely among them or using long feeding troughs, and each animal obtains a reasonable supply. If a competitive situation is set up by placing a single small pile of grain before such a group when hungry, an entirely different picture appears. One of the hens, which we will call the alpha or No. 1 animal, immediately approaches and feeds actively. Some other individuals may stay near the grain but make little or no effort to share in the food. If some of the grain becomes scattered, animals nearby may eagerly peck at it. Should the alpha animal notice them, it may make a threatening gesture or may peck the interloper. The latter immediately shows fear, retires, and never returns the peck. If the alpha animal is removed from the scene by the experimenter, another hen immediately takes her place and apparently by common consent is allowed to dominate the scene in the same way. Repeated removals of the top animal show that all the animals were eager to feed and were well aware of the food but were prevented from doing so by the individual at the food. This animal, called the dominant, was able to take precedence at the food and by its presence there prevented the other members of the group from even attempting to feed. Thus the group is seen to be composed of an array of individuals which form a hierarchy of precedence. Such precedence in competition assumed by an individual and acquiesced to by other members of the group is called "behavioral dominance." In stable chicken groups (Fig. 1.6) the order of dominance is usually clear-cut and forms a linear hierarchy in which alpha dominates all

| | Y | B | V | R | G | YY | BB | VV | RR | GG | YB | BR |
|---|---|---|---|---|---|---|---|---|---|---|---|---|
| Y | | | | | | | | | | | | |
| B | 22 | | | | | | | | | | | |
| V | 8 | 29 | | | | | | | | | | |
| R | 18 | 11 | 6 | | | | | | | | | |
| G | 11 | 21 | 11 | 12 | | | | | | | | |
| YY | 30 | 7 | 6 | 21 | 8 | | | | | | | |
| BB | 10 | 12 | 3 | 8 | 15 | 30 | | | | | | |
| VV | 12 | 17 | 27 | 6 | 3 | 19 | 8 | | | | | |
| RR | 17 | 26 | 12 | 11 | 10 | 17 | 3 | 13 | | | | |
| GG | 6 | 16 | 7 | 26 | 8 | 6 | 12 | 26 | 6 | | | |
| YB | 11 | 7 | 2 | 17 | 12 | 13 | 11 | 18 | 8 | 21 | | |
| BR | 21 | 6 | 16 | 3 | 15 | 8 | 12 | 20 | 12 | 6 | 27 | |

FIG. 1.6. Peck order as established by bringing twelve hens together is shown in the chart above. Each bird is designated by letters indicating the color code used to identify it. The number of times a given animal pecked other members of the flock is given in the vertical columns. The number of pecks and from which animals pecks were received are read in the horizontal column. It can be seen that each animal pecked only those below it in this liner hierarchy. (After A. M. Guhl, *Scientific American,* 1960.)

others; beta, all but alpha, and so on down the line, the omega animal being subordinate to the rest (Collias, 1944; Allee, 1951).

This pattern of precedence is found to apply to almost all competitive situations, at least in daylight when chickens are normally active. Thus the dominant animal assumes control of food, water, roosting places, choice of mate (by roosters, not hens), and any other competitive feature. The important point to recognize is that the precedence involved is not one which is fought over but is a regular characteristic mutually agreed upon and recognized by all members of the group. Occasionally a subordinate may overstep its bounds, and for such a social error it receives a severe peck. The subordinate does not peck back but flees. If in its flight it should encounter an animal subordinate to it, it may pass on the punishment by pecking. In a group of strange adult birds artificially brought together, this pattern of pecking is the most obvious expression of dominance, hence the term "peck order" often used by Schjelderup-Ebbe and later authors. Even aside from the fact that it is descriptive only of aggression in birds, this term is unfortunate, since it emphasizes the fighting that takes place. Such fighting, however, is largely an artifact resulting from the usual experimental conditions. We prefer the terms "rank-order" and "social dominance" for this discussion.

The induction of fighting by crowding enables the experimenter to determine dominance quickly, but from the theoretic viewpoint, the fighting involved is misleading. Natural or long-standing artificial groups that are stabilized and uncrowded show little actual fighting. The great importance of dominance in social life is that it acts as an organizing principle which minimizes aggression by, in effect, securing to the dominants the fruits of victory without disrupting group life by conflict.

Dominance behavior in chickens is clear-cut, vigorous, and quite uniform in expression, the alpha animal practically always dominating the rest. Such invariable dominance might be described as complete or absolute dominance; Schjelderup-Ebbe's term was "despotism." Allee used the term "peck-right." In other animals, however, dominance may be of a partial or relative character, called "peck-dominance" by Allee. In doves and pigeons, for example, no complete dominance of one animal over another appears; rather a count of pecks delivered and received may show a consistent balance in favor of one animal. On a statistical basis, pigeons in a flock may be arranged in a hierarchy of partial dominance, alpha delivering more pecks to beta or any of the others than he receives in turn and so on down the line.

Since partial dominance implies a failure of individuals to recognize one another and to accept social status once established, it is questionable whether it should be regarded as a part of social dominance in the sense of the term as used by Schjelderup-Ebbe. In any case, whereas absolute dominance is an organizing principle in social groups leading to a diminution of conflict within the group, partial dominance has little effect in this regard. It may be primarily a measure of the level of aggression in animals which do not form a well-organized group because of lack of facility in individual recognition.

Dominance hierarchies are not always linear. Especially when first formed they may be irregular, A dominating B, and B dominating C, but C dominating A. Such a triangular hierarchy tends quickly to break down into linear form in chickens but may persist for a long time in animals with partial dominance. Some animals may form hierarchies at two levels. All male chickens dominate all female chickens so that we may consider that there are separate male and female hierarchies. Prairie dogs similarly show a two-level arrangement. In mice under some conditions and in the Indian antelope in small herds,

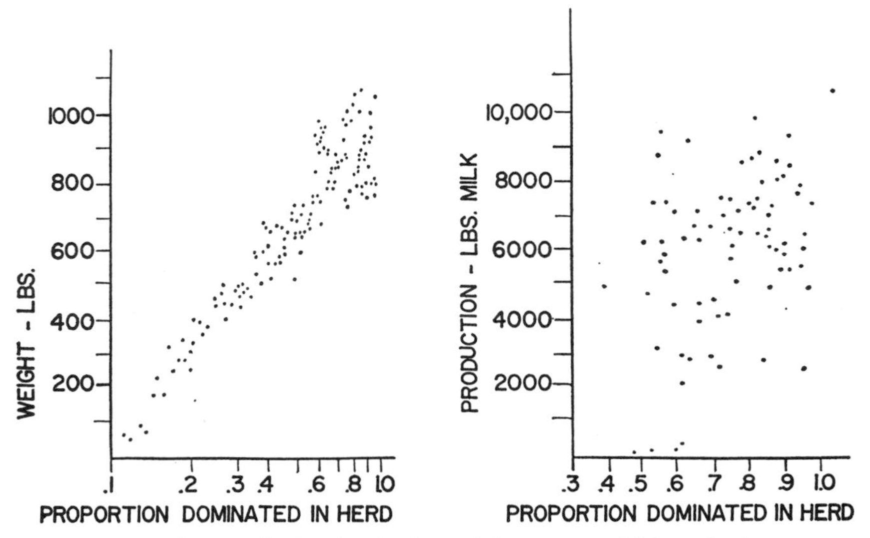


FIG. 1.7. Even such placid animals as dairy cows establish a dominance hierarchy in their groups. Dominance is established at about six months when the developing animals begin to take their pushing fights "seriously" after much play-fighting in younger animals. The "fights" are often mere threats by head movements or brief pushing so that these escape the notice of casual observers. Rarely are the animals so closely matched that real trials of strength occur. The plot on the left shows the high correlation between weight of animal and the proportion of the herd it dominates. Since weight correlates with age, the position of the cow in the hierarchy in these highly inbred animals seems determined largely by its birth order. As shown by the plot on the right, milk production does not correlate well with rank order. (After M. W. Schein and M. Fohrman, *Brit. J. Animal Behav.* 3 [1955].)

there is one dominant animal, all other members of the group being at one level in the hierarchy.

The vigor with which dominance is expressed may vary greatly in different species. Thus many herbivores may show only the mildest, if any, expressions of dominance in the grazing herd or may even feed from restricted bins without any evidence of precedence (Fig. 1.7). Yet in other species (the swordtail fish, ground birds like the chicken, carnivores such as the wolf, and some primates like the baboon and macaque monkey), dominance patterning of social relations is clearly important and strongly expressed, the dominant taking precedence in all competitive situations.

Besides species differences, another source of variation in the expression of dominance is that this behavior may be expressed in some aspects of life and not in others. Sexual behavior is most commonly the strongest area of the expression of dominance. Thus numerous ungulates (for instance, the red deer) show the strictest dominance among males in breeding season. The buck in rut collects and dominates a harem of does and rigorously excludes other males. Yet, out of rut during the major part of the year, the bucks live in loose herds showing little social interaction or expressions of dominance. Females, on the other hand, have well-organized, permanent herds showing clear-cut dominance patterning based on mother-offspring relationship with an old female as boss and leader of the herd. Almost no aggressive action need be shown in such a herd because each individual finds its place gradually as it grows up (Darling, 1937). Among Indian antelope studied in the zoo, there is one dominant male who asserts his dominance over other males conspicuously while on open range, but when the animals crowd around the feeding box, all evidence of his dominance disappears. He has to push his way into place as does any other animal. Thus, as in so many examples of animal studies, the behavior is "situation" specific. Dominance is commonly conspicuous in situations which are competitive in the normal life of the animal.

Dominance should not be confused with leadership. The lead in group movement often goes to the animals who are most alert to environmental change and, therefore, initiate the movement and are followed by the others. Where dominance is strongly and aggressively expressed, as in macaque monkeys, baboons, or the Indian antelope, the dominant males ordinarily do not maintain any sort of guard for the group and so do not initiate or lead. In the red deer, where the dominance of the older females is expressed more beneficiently, they maintain an alert and lead and direct herd movements. It is difficult to define the concept of leadership precisely enough for experimental analysis. In a simple situation Stewart and Scott (1947) found no correlation between dominance and leadership in a flock of goats.

An important aspect of dominance behavior in animal groups is the way it operates to make them closed societies. In a stable group, as natural groups generally are, each animal knows its status and keeps to it. If a stranger is introduced into such a group, the animal is evidently under considerable psychological disadvantage and is subject to persecution by all of the "natives." Thus in chickens, if a vigorous animal, say the alpha animal of one group, is introduced into another group, it first assumes a subordinate position, avoiding rather than retaliating attacks. However, as it becomes familiar with the situation, it may begin to fight back and gradually work its way up the scale to an appropriate position in the hierarchy. On the other hand, an animal not naturally so vigorous may be so beaten and oppressed as to succumb altogether with little show of resistance. Chicken farmers are familiar with this phenomenon and therefore will not ordinarily introduce a single individual into a strange flock. This rejection of the stranger (xenophobia) has been observed in numerous animals under field conditions. In the macaque monkey, for example, strangers, even females, attempting to enter a group are usually driven off by the males (Carpenter, 1942).

The establishment of dominance patterns under experimental conditions has often been studied to try to discover the factors involved (Collias, 1944; Guhl, 1953). When two strange hens are brought together, they immediately approach, watching each other closely. If there is a difference in size or vigor between them that is obvious enough to be detected by the human observer, the chickens can be counted upon to see it at a distance. In such a case one will, by lowering its head and avoiding the other, show that it accepts the subordinate position, whereas the other, holding its head high, shows its dominance by its freedom of movement. Again quoting Schjelderup-Ebbe (1935) for his colorful language describing his experience with numerous species of birds: "One bird, the subordinate, evinces apprehension, fear, and occasionally even terror of the other. The ruling despot discloses his identity by complete lack of fear toward the subordinate individual, and sometimes—strange to see—completely ignoring the other's existence."

If the birds being brought together are closely matched, they may approach, each one trying to keep its head higher than the other. If, by any accident such as stumbling, one head should be lowered, the issue is decided; that one becomes subordinate. In the usual experimental case, the animals are closely matched and approach close enough to fight. The fight is usually a matter of a few swift pecks, and one retires vanquished (Murchison, 1935). Such a decision is often seen under natural conditions where new relationships are being set up. In general, however, a fight occurs only when the opponents are so evenly matched that even to their expert eyes no other way of reaching a decision is apparent. In most cases the dominance-subordination relation is settled at the first encounter of the animals (Guhl, 1953). In many animals aggression takes the form of threat displays rarely involving real fighting (Fig. 1.8).

Once established, a dominance pattern tends to be highly stable. Some of the factors making for stability have been revealed by experimental and observational studies. For one, group-living animals have, in general, a rather good memory with respect to individual recognition within their own group. Thus a chicken removed for two weeks from a group can often be reintroduced without disturbance, since its status is remembered by all. After two weeks or a molt which

FIG. 1.8. This shows the fighting "dance" of two rattlesnakes. In their competitive aggressions these snakes attempt to push each other but do not strike. The fantastic figures assumed by such fighting snakes were once assumed to be courtship dances and are familiar to all in the caduceus symbol of the medical profession. (After H. Hediger, *Psychology of Animals in Zoos and Circuses*, 1955.)

changes their appearance, the recognition tends to be lost. The reputations of parrots and elephants for memory of individual associates, including human beings, extending over years seems to be well authenticated.

A second factor in stabilizing dominance organization is that the behavior of both dominants and subordinates tends to reinforce their status positions. Thus the dominant animal in most forms in which complete dominance prevails is readily recognized by the way it carries itself. As in the quotation from Schjelderup-Ebbe above, the bold look in the bird, the free movement, and indifference to the other animals characterize the dominant. Contrariwise, the manner in which the subordinate carries itself and avoids the dominant betrays its status. In some animals (for instance, the baboon and macaque) in which dominance is harshly expressed, each dominant male tends to occupy a semi-isolated area since other members of the group (except females in heat) tend to avoid it. The posturing "advertisement" of dominance in the Indian antelope is carried to an extreme. He assumes an awkward, stiff, strutting walk, holding his head high on stiffly bent neck, with his ears folded tightly flat against the neck. In many familiar species of deer and in many other large mammals, the rutting male shows similar awkward poses. Clearly then, many dominants maintain their position by a special conspicuous carriage, which seems to gain in conspicuousness by its non-adaptive character. Thus the folded ear of the antelope does not point toward the source of a sound and is in marked contrast to the mobility of the ears of the other animals in the group.

The importance of this factor of display is shown by the phenomenon of prestige as a factor maintaining social status. The dominant hen, for example, maintains its position of dominance as it gets old, weak, and less agile, even while losing its vision. It may then make even more of a show of its dominance display, carrying it so far as to seek occasion to chase or peck at subordinates. Eventually, however, one or more of the subordinates, despite its conditioning to subordination, begins to show signs of revolt. At first, perhaps, the revolt is shown only by making pecking and threat gestures in the direction of the dominant when the latter is turned away or is at a safe distance. Gradually the boldness of the subordinate increases until it eventuates in the inevitable battle and quick defeat of the former despot. An old and infirm despot when deposed in this way is indeed a sorry sight, becoming utterly dejected and full of fear. In a chicken yard such an animal often dies within a few days, presumably as a result of the traumatic experience as well as physical injury inflicted by its former subordinates. The maintenance of social status by prestige as well as the extreme effects that follow its loss shows how important a role dominant-subordinate relations play in the lives of at least some social animals.

It would appear evident that since a stable dominance pattern reduces the amount of fighting that takes place in an animal group, it must help to make the group life more efficient. Such experimentation as has been done tends to support that concept. For example, a comparison was made between two groups of hens allowed to stabilize their social structure and two in which individuals were being interchanged frequently so that the social structure never could settle down (Guhl, 1953). The stable groups pecked less, ate more, gained more weight, and laid more eggs than did the others. When the welfare of the individual is considered, it is also found that, in flocks kept on short rations, the dominants secured more food than did the subordinates; indeed, top dominants gained weight while subordinates starved. It would seem that not only may dominance lead to survival for the individual under conditions of stress, but it may also help in group survival. It is well known that among ungulates such as deer and bison a great increase in population may lead to overgrazing of the area in which they live. Then if a particularly hard winter comes along, the entire herd may be lost because of the lack of food, since such grazing animals lack any well-developed dominance organization with respect to food competition. However, in animals with dominance in feeding, the animals higher in the scale secure all available food, whereas the subordinates are starved out early. Since environmental stresses such as food supply and weather are frequently controlling factors in the survival of animals, it is evident that the advantage enjoyed by dominants enables them and, with them, the species to survive. It would appear, therefore, that there are strong evolutionary pressures making for the development of dominance as a principle of social organization. Whereas it is certainly not universal among group-living animals and in some species takes exceedingly mild and limited forms, it is a principle which nevertheless helps stabilize many types of social groups. At the same time, by eliminating many of the harmful effects of aggressive activity expressed within a group, it permits evolutionary pressures to push the development of the capacity for aggression to a high degree. Thus dominance patterning of social organization is clearest in those group-living animals which show a high level of individual aggression.

Dominance hierarchy is by no means the only factor in the control and modification of aggression in the social life of animals. In later chapters we will discuss the role of courtship, parental care, early experience, and learning in relation to aggression. At present, however, we wish to turn to territoriality because this phenomenon is very closely connected with dominance in control of aggression in animal societies.

### Territoriality

In an attempt to see this phenomenon in its most general aspects, we shall define territoriality as any behavior on the part of an animal which tends to confine the movements of the animal to a particular locality. This definition differs from those conventionally used for reasons which will become apparent later in the discussion.

Territoriality is a phenomenon of overriding importance in the lives of some animals, particularly birds. Although occasionally noted by ornithologists earlier, it was not until 1920, when the English student of birds Eliot Howard published his book on territory in bird life, that the significance of the phenomenon was really appreciated. Perhaps the most famous American study is that Margaret Nice made of the song sparrows in the Cleveland area (Nice, 1937). From this study we take the following account which illustrates territorial behavior in a representative song bird.

About half of the males and a small proportion of the females remain in residence during the winter, the others migrating south. Each resident bird tends to confine his activities to a restricted region but, particularly in bad weather, may join with others to form a winter flock. In the early spring, often as early as January if there is a warm spell, each resident male begins to show special behavior in relation to its particular area, a plot usually of about one acre in extent. Here he takes up a conspicuous place on a reed or tree and sings loudly and persistently. This behavior becomes more and more marked as the season progresses until in March the animal stays permanently in its own territory and sings frequently from his "headquarters" or display station. Should another song sparrow alight anywhere near, the resident assumes a very watchful and aggressive attitude which is usually sufficient to drive off the intruder. In some cases the intruder, usually a migrant arriving from the south, stands his ground and shows that he intends to stay by remaining in spite of the threats of the territory holder. In this case a rather definite behavioral interchange or "territory establishment ceremony" ensues. The newcomer puffs out his feathers, often holds one wing fluttering straight up, and sings softly. He appears subordinate to the confident dominant resident. The latter, now silent, watches him closely for a while as he flies from bush to bush. Then he gives chase. The intruder flies off but persistently returns and eventually turns on the resident, and a fierce pecking fight on the ground ensues. The battle is brief; the intruder, if defeated, flies away. If, however, he holds his own, the birds separate. Each, retiring to a different high point, sings loudly. It is thereafter noted that the original territory of the resident is now divided into two territories. Each bird keeps strictly to his own territory, clearly recognizing and respecting the invisible boundaries separating it from that of its neighbors. Each bird frequently sings from his display area in his territory and behaves as a self-confident dominant, driving off any small bird transients and fighting any male song sparrow intruders. When the females arrive from the south, pairing takes place in the male's territory, and the male stops singing for a month or two. The nest is built there, and both birds largely confine their activities to this region. The male remains active in defending his territory; although after the spring migration is over, there is much less occasion for fighting and territorial defense behavior slackens.

The above example is typical of territoriality in its classic form

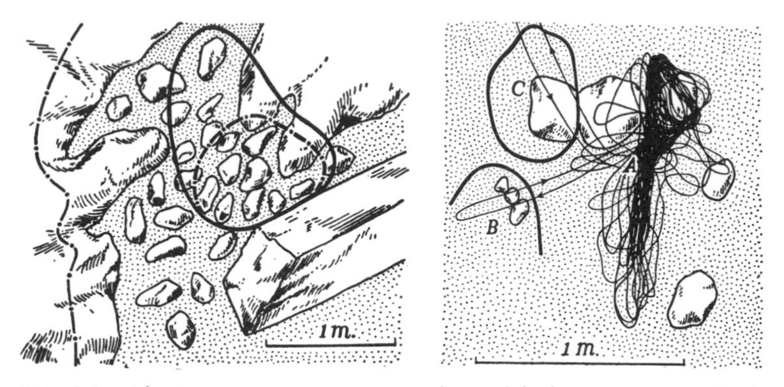


FIG. 1.9. The Japanese ayu is a stream-living fish that maintains a feeding territory usually around selected rocks in the stream bed (*left*), the central area of which it defends from other ayu. When crowded, those animals, unable to maintain a territory, live in non-territorial aggregations. At right is a tracing of three minutes of an ayu's life showing how its movements are confined to its territory except for two excursions made in driving off interlopers. (From Miyadi, *in* A. A. Buzzati-Traverso (ed.), *Perspectives in Marine Biology,* 1958.)

particularly as seen in song birds. It is a geographical area to which the animal confines itself and from which it excludes others, particularly members of the same species, except its mate; such a territory is a "defended area." Such defended territories are of widespread occurrence among vertebrates (Fig. 1.9), being especially common in bony fish, reptiles, birds, and some orders of mammals, particularly primates. Many students of animal behavior use the word "territory" to refer only to this type of defended area (Collias, 1951; Burt, 1943; Carpenter, 1958).

On the other hand, the locality sense of animals may take a variety of forms, some of which bear resemblance to this classic type but differ in that the animal shows no tendency to exclude others of the species from it. Thus many mammals, such as rats, mice, pronghorn antelope, and many deer, show a very strong predilection for remaining within a limited area with which they are evidently familiar (Fig. 1.10). Thus field mice in nature and rats in warehouses have been found to confine their activities to very limited space of a few hundred square yards or a few rooms. Hunters chasing pronghorn antelope in

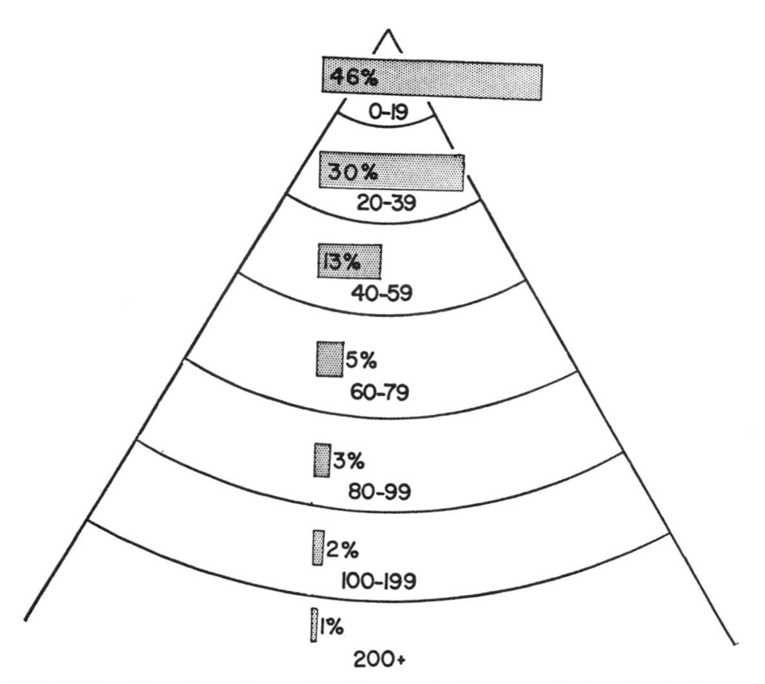


FIG. 1.10. The chart shows the distance in feet at which individual rats have been found (recaptured) after being released from the first point of capture in an urban site. This illustrates the strong tendency of rats to remain in the same area (home range); 89 per cent of the captures were made within 60 feet of the release point. (After D. Davis, *Quart. Rev. Biol.*, 28 [1953].)

our western plains often noted that it is impossible to drive an individual outside of the few square miles of area in which it is found. Even at considerable risk the animal persists in circling back to keep near to home. Such a sense of locality, which is characterized by the positive element of attraction to the area without the negative element of driving others off, may be called home range or home-range territory. Unfortunately, the distinction between defended and home-range territory cannot always be maintained in practice, since our knowledge of the natural behavior of animals under the varied natural conditions of their lives is often insufficient to enable us to say to what extent others are excluded from the territory. The unqualified term "territory" is therefore useful for such cases, and it can readily be qualified as home-range or defended as our knowledge justifies.

## FORMS OF TERRITORIALITY

At best, the distinction between home-range and defended territory is not an absolute one, since it depends upon the very variable factor of aggressiveness shown toward others. Such aggressiveness fluctuates not only from species to species but within the same species from time to time and occasion to occasion. Thus, in the song sparrow discussed above, the males that remain in residence throughout the year tend to stay in the neighborhood of their breeding territory at all seasons. Yet in the late fall and early winter, they do not exclude other song sparrows from the area. In fact, the tendency toward exclusion of others is strongly expressed only in the early part of the breeding season, beginning usually in January or February. In June it wanes distinctly. During the molting period in September, the animal is very retiring and does not challenge trespassers. Later he may challenge adult males again to some extent but permit young males of the current year's brood to remain as long as they vocalize only with juvenile call notes and do not attempt adult song. In December the resident male neither sings nor challenges any others. Thus the male song sparrow maintains defended territory early in the breeding season and home range later.

Territorial behavior may be shown by a group, such as a herd or flock, rather than by an individual or pair. Such group territories are known to exist for many mammals and some birds. The red deer of Europe, for example, live in herds (Darling, 1937). Except in the rutting season, the males live in loosely formed herds separated from the better organized herds of females with their young. During rut, the adult males abandon their own territories and wander far and wide through the territories of female herds, seeking to round up as many females as possible and keep them together in a small group. The male rigorously excludes all other adult males from this harem area by constant patrolling and fighting when necessary. The hinds, however, do not share the stag's rut territory but tend to wander off unless persistently rounded up by the males. As a male is exhausted, he loses control of his hinds and returns to the uplands for rest. Territoriality in the red deer is thus complex, at one time being a group phenomenon, at another, an individual one confined to the male sex. Group territory seems to be of the home-range type. The rut territory is not a fixed geographical area but simply the neighborhood around the hinds as they move about grazing. Game cocks in a farm yard may similarly show a mobile territory, each cock staying with his hens and each such group avoiding the others so that there is no conflict. Yet each of these expressions shows a special behavioral relation the animals have to a special area.

Functionally, territoriality may be related to one or many of the fundamental life processes—reproduction, feeding, etc. As was obvious in the examples previously discussed, the typical defended territory seems to function primarily in relation to reproduction. Thus territoriality among song birds is most evident during the period of nesting. In an animal such as the song sparrow, since the mated pair remain within the territory during this time, it also serves for feeding, nesting, roosting, and raising the young. Such a territory may be characterized as a feeding-breeding territory. This, however, is not the case in other species. Many marine birds, for example, range over the ocean for feeding. Their nests are usually in great colonies along a favorable shore. In these colonies, territoriality is expressed very strongly with regard to a small area around the nest. Thus gulls, terns, gannets, and many other marine birds nest in large colonies in which each pair has a small nesting area, sometimes as little as a square yard or so, which is vigorously and constantly defended against trespassers, even young chicks from neighboring nests being excluded (Armstrong, 1947). Herring gulls pair up before adopting such breeding territories but conduct all their reproductive activities thereafter in the territory and desert it only when the young leave (Tinbergen, 1953). In other birds territoriality may function during only part of the reproductive process. Some male ducks, for example, maintain an out-of-the-way territory to which they conduct the female for copulation. A most interesting form of mating territory is lek formation, as shown by the European ruff, our prairie chickens, grouse, and others. In this type large numbers of males establish small individual mating territories in close proximity. The European ruff stands in his territory, watching to see that no other male oversteps its borders. When a female (reeve) appears, the males go into a frenzy of display, throwing the collar of long colorful feathers up around their neck to produce the conspicuous ruff and taking stiff "ecstatic" poses. The female chooses her mate without interference, and once she commits herself, the other males lose all interest in her movements (Armstrong, 1947). Members of a sage grouse lek not only display themselves visually but make their characteristic loud, booming sound. Since in these cases the female departs after mating and raises the young by herself, the territory is functionally restricted to mating. On the other hand, the winter territory of such birds as the mocking bird and English robin serves only in relation to self-maintainance and not reproduction (Lack, 1943). Group roosting territories are a familiar sight in winter for birds such as starlings. Bat caves may also be considered roosting territories, though, of course, mating also takes place there.

An interesting analogy to lek formation occurs in some fish, mostly stream-living. In the rainbow darter, for example, the breeding males congregate in shallow, gravelly places. Here the larger males stake out small breeding territories from which they drive all other males, maintaining a continuous guard. Females are permitted to enter, copulation and spawning taking place in the territory. In the red-bellied minnow, males congregate in a part of the stream appropriate for spawning. When a female swims in, males congregate around her, usually one on each side. Males compete for position around the female, but there is no fighting; and once two animals are lined up on either side, they are not displaced. Thus, in spite of the fact that many males are in the area, aggression is minimized, and two males co-operate from either side in fertilization.

### RELATION TO AGGRESSIVE BEHAVIOR

A most significant aspect of territorial behavior from our present point of view is its relationship to aggressive behavior. The central point, of course, is that a territory of the defended type is maintained by the aggressive actions of the territory holder. Such action is directed essentially at "competitors," most often sexual competitors of the same species. Though the decision in a conflict over territory may involve actual fighting, even fatal combat, it is apparent that, once decided, the isolation of the aggressive individuals in separate territories eliminates further fighting. Thus territoriality minimizes the amount of actual harm done to members of the species by fighting. All types of aggressive behavior are not necessarily eliminated by territorial isolation. Many territory owners patrol the borders of their territories, singing persistently or otherwise displaying aggressively. Bulls of the fur seal and related species, for example, repeatedly charge to the edge of their breeding territories, making formalized open-mouthed thrusts at neighbors who reciprocate. Occasionally the charge takes the form of a belly slide that ends with the two animals snout-to-snout at the territory boundary, each "puffing at the other like a small locomotive" (Bartholomew, 1953) (Fig. 1.11). The absence of real fighting in the stabilized territory permits the reproductive life of a mated pair of song birds, for example, to be carried through without interruption by competing members of the species. In other animals, the fur seals

FIG. 1.11. The defense of territory by bull sea elephants involves loud trumpeting (the enlarged proboscis acts as a resonating chamber) and labored charging at rivals, each animal going only to the border of his territory. Thus neighboring males frequently end their charges in the upright posture as shown above. (From a photograph taken in the Antarctic by A. Saunders.)

for instance, life is not so quiet, since young bulls are more persistent in attempting to reach the females throughout the season.

Careful watch of the behavior of the territory defenders reveals that the relative tranquillity that often exists is dependent upon an important relationship between territoriality and dominance behavior. The well-established territory owner has an enormous advantage in any conflict with intruders. It can often be observed that a bird within its territory behaves and carries itself with the characteristic self-assurance of a dominant. But when for any reason it leaves its territory and trespasses on that of its neighbor, it assumes the lean look and furtive behavior of a subordinate. This is most dramatically seen in some fish which show their aggressive state by their pigmentation. Some male cichlids, for example, as they swim away from their territory, blanch noticeably in their nuptial colors. In many animals there seems to be a rough proportionality between the "self-confidence"—or tendency to dominate—and distance from the territory headquarters. Thus, when territorial manakins were experimentally subjected to competition for a stuffed female set down in their territories, each animal's courtship was expressed with a vigor that varied in proportion to the distance of the lure from the center of its territory. Similar results were seen with great tits, where the degree of dominance at an outside feeding station was found to vary with the distance from the animal's territory. This tendency for the animal to show dominance in relation to the site of the conflict in respect to its territory is seen in naturalistic observations of many animals. For example, in territory establishment in the song sparrow and in many territorial species of fish, reptiles, and mammals, intruders have difficulties in proportion to the smallness of the territorial possessions into which they attempt to insert themselves. This phenomenon generally places a limit on the reduction in size of territories; for when an owner's territory becomes small enough, he will fight to the death to prevent loss. This limitation of compressibility of defended territory generally has important ecological or practical consequences, as we shall see shortly. Here, however, we should note the resemblance to the behavior of a dominant in a group, who so often maintains a characteristic dominance isolation, as we mentioned above for baboons. An analogous phenomenon is the formation, within large herds, of subgroups around each male, as described for game cocks. From this point of view, territoriality may be viewed as only one, though perhaps the most definitive, of the ways in which a dominant tends to isolate itself or be isolated by the creation of a "no man's land" around it (see Chap. 10). Hediger has emphasized the importance of such "flight" distance relations between trainer and animal in circuses and zoos in maintaining the appropriate dominance relations between them (Hediger, 1955).

As much of territoriality is related to mating behavior, we often find that territorial defense is the function of the male only. This is particularly true of fish, like the sticklebacks, that build nests and of the lek birds. But in song birds, where male and female remain within the territory, the female may in the course of a few days gradually assume some defense activity, either against all intruders, as do the English robins, or against females only, as do some ducks and others. In some cases the roles are reversed; the females of the red-necked phalarope, for instance, hold the territory. The defense of lairs as territories by female mammals is common. The European rabbit defends the area of her nest irrespective of the presence of young. Within a male chameleon's territory several females may defend subterritories against one another (Greenberg and Noble, 1944). Winter territories may be held by female song birds such as the California mocking bird or the English robin. Thus, although defended territoriality is predominantly a male characteristic, it is by no means exclusively so.

### THE ECOLOGICAL SIGNIFICANCE OF TERRITORY

As biologists, we are inclined to assume that any basic characteristic of organisms, behavioral or otherwise, is likely to have survival value. Hence, Howard, in his original discussion of territoriality, attempted to analyze the ecological value of territory holding. Many of the benefits seem obvious; yet since experimental work in this area is extremely difficult, it is impossible to be sure to what extent benefits that seem reasonable are really of importance. We must depend chiefly on an analysis of observations, especially comparative observations, of different species or the same species under differing conditions as a basis for our interpretations. Though some points of Howard's analysis have been strongly disputed—particularly the value of territory as an exclusive feeding range—for the most part his interpretation has stood up well and will be followed here.

We may consider first the relation of territory to reproduction. The male bird establishing his territory in the spring makes himself conspicuous. Song birds do this by loud and frequent singing, usually from conspicuous perches. In some species, the sky lark, for example, the male indulges in conspicuous flight over the area; in others, awkward poses are assumed, such as that of the European stork sitting on his nest klappering and waiting for a passing female to fly down to him. This self-advertisement presumably helps the female find unmated males. This is further indicated by the commonly observed fact that the males in many species become much quieter after mating. In lek birds, the mass effect of the advertisement in aggregation combined with the use of the same areas year after year likewise facilitates the finding of mates by females when they are ready. Though matchmaking may be important in some instances, however, it must clearly be a secondary consideration in typical song bird territory, since territory defense persists after mating, and some birds, for instance, herring gulls, pair before territory formation.

The persistence of territorial behavior beyond mating suggests that it is helpful in maintaining the association of the mates throughout the reproductive period. As we shall see in more detail in the discussion of reproductive behaviors, the co-ordination of the behavior of both mates during the long and complex sequence of activities that are necessary for effective bird reproduction is a major problem. By confining the reproductive individuals to a small territory, their behavioral and physiological co-ordination is facilitated. This would appear to be one of the most general, as well as important, functions of territoriality. Such facilitation may extend through to include parent-offspring co-operation. We shall discuss it further in Chapter 4 but here one aspect of co-ordination related to dominance ought to be mentioned. Dominants often interfere with mating among the subordinates in the group. Under experimental conditions, it is often found that subordinates among dogs and ducks and fowl often make no attempt to mate when in the presence of the dominant, although proving eager when the dominant is removed. In some species, such as the chicken and dog, where this suppression of mating is well marked and persists even in the absence of the dominant, the phenomenon has been called "psychological castration" (Guhl, 1953). In other instances (e.g., seals and rhesus monkeys), subordinate males in the group achieve mating only when the dominant is completely occupied by other females. Thus, by isolating each pair in an area in which they are effectively dominant, territoriality favors successful mating.

Another function which Howard saw in territoriality is that of maintaining an economic balance between food and population. In the song bird territory, he saw the establishment of freeholds wherein each family would find the food resources necessary for survival. Young birds have enormous appetites and high growth rates; nestling song sparrows increase their weight more than tenfold in the first ten days. At the same time, nestlings are delicate and must be brooded to be kept warm and protected from the weather. The presence of an uncontested food supply nearby is of obvious value. Yet the importance of this item, even in song bird territory, has been disputed. Food competitors of other species, for example, are not attacked by most species, and the intensity of territorial defense often drops markedly before the young are fledged, even though their food requirements are then at the highest point.

Even though the idea of territory as the freehold of an individual family does not seem to be as important as Howard thought, territoriality still seems to play a role in regulating animal economics in another way, namely, in acting as a population-regulating mechanism. Since territories are not indefinitely compressible, it is clear that, as population goes up, the range of the species is extended by the new individuals claiming territories peripheral to the group. In general, this entails less favorable areas. Song bird males that are unable to find adequate territories are generally unable to breed at all. Thus the population in a given area would tend to be stabilized around the number of families that can be supported by the resources of the area for that species. Some evidence that this is actually operative is seen in the fact that in some species the removal of one member of a territorial pair is followed, sometimes within a day or two, by the settling in its place of another bird (Lack, 1943). Such replacements presumably had been part of an excess of non-territorial, non-breeding birds. This phenomenon of ready replacement is by no means universal in song birds, and it may well be that in some species—such as the song sparrow, where it was not found—other factors may have been regulating population size. Perhaps, too, it varies with the year or the place. Nevertheless defended territoriality does seem to place an upper limit on increases in population. As a consequence, in species with this type of territory we do not find the tremendous variability in numbers from one year to another that is found in some other animals. In mice and rabbits, for example, which show home-range behavior, a tendency for population to run in cycles in some areas has been observed. The population builds up to a very high density in favorable years only to be cut down suddenly to a very low level. The lemming of Norway, a mouse-like animal, is the classic example of an animal whose population builds up to high density and produces a plague of animals. These move out of the mountains and keep migrating, eating farm produce, swimming rivers in their path, and pushing on until finally they reach the ocean into which they plunge to their death. Such cycles are well known among the Arctic rabbits of Canada, the lynx, and other predators that depend upon them. Such animal plagues of mice and rabbits, unlike the lemming plagues, often end with a great dying-off whose cause is obscure (see also Chap. 3). Defended territoriality protects a species from such violent population fluctuations and illustrates the importance of the type of territorial behavior for an animal's general ecology.

## Summary

We may summarize the conclusions we have arrived at in this chapter as follows:

1. Modern biology supports the concept that the primary driving force making for evolutionary change is natural selection. Therefore, all characteristics, behavioral as well as anatomical and physiological, must generally be adaptive—enabling the animal better to survive.

2. Social organization is often of great survival value for the group. Even at the most primitive level of loose aggregation, sociality may have decided advantages for survival.

3. However, evolutionary processes must be expected, at least in vertebrates, to favor competitive and aggressive behaviors on the part of the individuals. These would tend to disrupt social life. Since, in spite of this, group formation in vertebrates is very common, we are led to expect that there must be ways in which aggressive behavior is kept under sufficient control to prevent its interference with sociality.

4. Two factors which operate very widely in vertebrate social groups to control aggressive behavior are (*a*) dominance hierarchy based upon individual recognition among the animals of a group and (*b*) territoriality which confines aggressive and other behaviors to limited areas. These factors and their importance in the organization of social groups in vertebrates are described in some detail in this chapter.

## REFERENCES

Allee, W. C. 1951. *Co-operation among Animals.* New York: Henry Schuman.

Alverdes, F. 1927. *Social Life in the Animal World.* New York: Harcourt Brace & Co.

Armstrong, E. 1947. *Bird Display and Behavior.* New York: Oxford University Press.

Bartholomew, G. 1953. Behavioral factors affecting social structure in the Alaska fur seal. *In:* J. B. Trefethen (ed.), *Trans. Eighteenth No. Amer. Wildlife Conference.* Washington, D.C.: Wildlife Management Institute.

Burt, W. 1943. Territoriality and home range concepts as applied to mammals. *J. Mammal.,* **24:** 346–52.

Carpenter, C. R. 1942. Characteristics of social behavior in non-human primates. *Trans. N.Y. Acad. Sci.,* Ser. II., **4:** 248–58.

———. 1958. Territoriality. *In:* A. Roe and G. G. Simpson (eds.), *Behavior and Evolution.* New Haven: Yale University Press.

COLLIAS, N. E. 1944. Aggressive behavior among vertebrate animals. *Physiol. Zool.*, **17**: 83–123.

———. 1951. Problems and principles of animal sociology. *In:* C. STONE (ed.), *Comparative Psychology*. New York: Prentice-Hall.

CROOK, S. 1961. The basis of flock organization in birds. *In:* W. H. THORPE and O. L. ZANGWILL (eds.), *Current Problems in Animal Behaviour.* Cambridge: Cambridge University Press.

DARLING, F. F. 1937. *A Herd of Red Deer.* London: Oxford University Press.

DOBZHANSKY, T. 1951. *Genetics and the Origin of Species.* New York: Columbia University Press.

GREENBERG, B., and NOBLE, G. K. 1944. Social behaviour of the American chameleon (*Anolis carolinensis* Voigt). *Physiol. Zool.*, **17**: 392–439.

GUHL, A. M. 1953. *Social Behavior of the Domestic Fowl* (*Tech. Bull. 73*). Kansas State College, Manhattan, Kans.: Agriculture Experiment Station.

HEDIGER, H. 1955. *Psychology of Animals in Zoos and Circuses.* New York: Criterion Books.

HOFSTADTER, R. 1944, 1955. *Social Darwinism in American Thought.* Boston: Beacon Press.

HOWARD, E. 1920. *Territory in Bird Life.* London: John Murray. Reprinted, 1948; London: William Collins Sons & Co.

KROPOTKIN, P. 1914. *Mutual Aid.* New York: Alfred A. Knopf.

LACK, D. 1943. *The Life of the Robin.* Reprinted, 1953; London: Penguin Books.

MURCHISON, C. 1935. The experimental measurement of a social hierarchy in *Gallus domesticus,* I. *J. Gen. Psychol.,* **12**: 3–39.

NICE, M. 1937. Studies in the life history of the song sparrow, I. *Trans. Linnaean Soc.* (N.Y.), **4**: 1–247.

SCHJELDERUP-EBBE, T. 1935. Social behavior in birds. *In:* C. MURCHISON (ed.), *Handbook of Social Psychology.* Worcester, Mass.: Clark University Press.

SHAW, E. 1962. The schooling of fishes. *Sci. Amer.*, June, 1962.

STEWART, J., and SCOTT, J. P. 1947. Lack of correlation between leadership and dominance relationships in a herd of goats. *J. Comp. Physiol. Psychol.*, **40**: 255–64.

TINBERGEN, N. 1953. *The Herring Gull's World.* London: William Collins Sons & Co.

WILLIAM ETKIN

# 2

# Neuroendocrine Correlation in Vertebrates

## Introduction

In this chapter we shall discuss aspects of the nervous and endocrine systems which will be directly useful as background information in reading the subsequent chapters, particularly those dealing with experimental analyses. We shall assume that the reader is familiar with the elementary physiology of these systems as conventionally discussed in college biology or psychology texts and shall stress those topics in which newer knowledge of the nervous or endocrine system has contributed insights relevant to the modern analysis of behavior.

## General Functions of the Pituitary

The pituitary, or hypophysis, is a double organ, part glandular (the adenohypophysis) and part nervous (the neurohypophysis). There are seven well-recognized hormones produced by the adenohypophysis, six by the cells of the largest part, the anterior lobe, and one by the cells of the intermediate lobe, which is closely attached to the neurohypophysis. The primary site of action of two of the hormones of the adenohypophysis is upon non-endocrine tissues of the body, whereas the other five, called tropic hormones, act chiefly upon other endocrine glands, whose activity they control. Because of its control of other endocrine glands, the pituitary is popularly called the "master gland" of the endocrine system. We may first consider the two directly acting hormones.

## Animal Pigmentation

The hormone produced by the pars intermedia is called intermedin, or the melanophore-stimulating hormone (MSH). Its only known function is to help regulate pigmentation. In some fishes, amphibians, and reptiles this hormone acts upon the cells bearing black pigment. The pigment is melanin, and the cells are called melanophores. Intermedin induces the spreading of the pigment within the melanophores and thus darkens the animal's skin. This hormone is part of the mechanism by which these animals are able to modify their body color to match and blend with that of their environment. It is believed by some to play a role in skin darkening in mammals as well.

While we are discussing the role of intermedin in pigment regulation, it might be well to mention two non-pituitary products which are also believed by some investigators to play a role in such regulation—adrenalin, secreted by the adrenal medulla and melatonin, a substance extractable from the pineal region of the brain. These act in opposition to intermedin in that they induce a concentration of the pigment in the melanophores. In many animals there are cells bearing pigments of other colors, such as red, white, etc., which are likewise sensitive to these hormones. It seems probable that all three hormones are involved in the alterations of an animal's color under varying circumstances of background adaptation and physiological states, such as excitement. We know that other hormones, such as the sex hormones, may act upon an animal's pigmentation in particular cases. The hormones concerned in control of animal pigmentation thus vary from species to species.

## Growth Hormone

A second non-tropic hormone, this one produced by the anterior lobe, is called the growth hormone, or somatotropin (STH). Its most diagnostic effect is to promote growth in hypophysectomized mammals, but it appears to act on many metabolic processes in the body, such as sugar and protein metabolism.

## Thyroid Function

The pituitary controls the rate of secretion of the thyroid gland by its thyroid-stimulating hormone (TSH), or thyrotropin, secreted by the anterior lobe. In warm-blooded animals (birds and mammals) the principal hormone of the thyroid gland, thyroxin, acts to stimulate the general chemical activities of the body and thus produces a rise in basal metabolic rate. In this way it assists in maintaining the constant body temperature in these animals and probably plays a significant role in seasonal adjustments to temperature change. In the absence of TSH the thyroid gland maintains a low rate of thyroxin production. Variations in the activity of the thyroid are dependent upon pituitary control, exerted by way of variations in the rate of TSH production. The interrelations of the two glands, the so-called pituitary-thyroid axis, is stabilized by a "feedback" mechanism whereby the amount of thyroid hormone in the blood regulates the secretion of TSH by the pituitary. The feedback arrangement tends to keep the axis upon an even level of activity since the inhibitory feedback effect of thyroxin cuts down excess activity of the pituitary. The system is so much like a thermostat arrangement for regulating temperature that it has been spoken of as the "thyrostat." Different levels of thyroid activity in relation to different needs of the organism are regulated by the level at which the cutoff point of the thyrostat is set, just as the temperature of a room is regulated by the setting of the thermostat.

## Adrenal Function

A second tropic hormone of the anterior pituitary is the adrenocorticotropic hormone (ACTH). This plays a role in regulating some (but not all) of the secretions of the adrenal cortex, similar to the role of TSH in thyroid regulation. As in the latter, the pituitary-adrenocortical axis is stabilized by a feedback relation between the two glands.

The adrenal cortex is, however, much more complex than the thyroid. It produces many hormones, though all are of a similar chemical (steroid) nature and may be referred to as corticosteroids. Some of these steroids act principally to control metabolic activities related to various foodstuffs, particularly glucose and are therefore called glucocorticoids. It is these which are chiefly controlled by pituitary ACTH. Other corticosteroids, called mineralocorticoids, are more closely involved in the regulation of the mineral-salt and water balance. These are produced by the outer layers of the adrenal cortex and appear to be independent of the pituitary but to be under the influence of hormones produced by the pineal gland and adjacent areas of the brain.

The central region of the adrenal is known as the adrenal medulla. This produces two hormones known as adrenalin and nor-adrenalin, sometimes referred to collectively as the catechol amines. These have long been known to act physiologically in a manner resembling the sympathetic nervous system (sympatheticomimetic) in the regulation of internal organs (heart, digestive tract) and various metabolic processes. These may be designated as vegetative functions, as distinct from the activities of the skeleto-muscular system by which movement, designated as somatic activity, is produced.

The sympatheticomimetic functions of the adrenal medulla are mediated through the catechol amines which the sympathetic nerve endings (with few exceptions) secrete at their terminals and which affect the immediately surrounding tissues. Thus the adrenal medulla and the sympathetic nerves use the same chemical substances to transmit their stimuli, the medulla pouring these substances into the general circulation where they act diffusely on the entire body over a period of hours. The same transmitter substances at the sympathetic nerve terminals act locally and persist only for seconds, since enzymes in the tissues quickly destroy them. The cells of the adrenal medulla arise during development from the same embryonic primordium (the neural crest) that gives rise to the cells of the sympathetic nervous system, so that, from an anatomical point of view, the medullary cells may be regarded as sympathetic cells that have lost their axons and become specialized in producing large quantities of the transmitter substances to pour into the blood stream.

The adrenal cortex, on the other hand, appears to be anatomically quite distinct from the medulla. It arises from a different primordium and, in the lowest vertebrates (sharks), is entirely separate from the medulla. In the amphibia and all higher vertebrates these two structures are intimately fused. In the mammals the cortical cells form a separate outer layer called the cortex, but in other vertebrates they are scattered through the medullary tissue as islands of cortex. It is not clear from a physiological viewpoint why the two parts need to be so close together, but functionally, and particularly from the point of view of behavior, the activities of the entire complex of sympathetic, medullary, and cortical systems are interrelated. Their functional importance in behavior can be understood on the basis of two theories which are widely, if not universally, accepted by physiologists.

The first of these is the emergency theory of the function of the adrenal medulla, which we may view here in the context of the general function of the catechol amines. The sympathetic system functions in facilitating bodily adjustments to various environmental changes by inducing compensatory shifts in tissue activity. For example, at the initiation of active exercise, heart action is accelerated and the circulatory system speeded up in other ways by the sympathetic nerve stimulation. When the animal faces a sudden emergency requiring widespread alterations of tissue activity, the entire sympathetic nervous system is thrown into immediate activity. Such general sympathetic stimulation results in a complex group of responses which, as Cannon—the physiologist who developed this theory—pointed out, constitutes a pattern of changes which enables the organism to meet the environmental challenge. As examples: the sympathetic activity increases the release of sugar by the liver, increases heart rate and volume of blood delivered per beat, shifts blood circulation from the digestive tract to the muscles. Such sympathetic action is, however, of short duration, since, as stated above, the catechol amines secreted at the nerve endings are quickly destroyed by tissue enzymes. In such an emergency the adrenal medulla is also activated by its nerves and secretes large quantities of catechol amines into the blood stream. As a result, the same pattern of tissue activity set in motion by the sympathetic nerves is continued by the activity of the adrenal medulla. Thus, according to this theory, the medullary tissue functions as a mechanism to meet emergencies by extending the sympathetic response throughout the body and over longer periods.

A parallel theory regarding the functional importance of the adrenal cortex was developed by Selye and has found extensive application in relation to behavior, as we shall see in Chapter 3. According to this concept, the various metabolic and other changes induced by the action of the adrenocortical hormones, especially the glucocorticoids, mobilize body resources to meet long-term stresses, such as major injuries, exposure to cold, etc. By increasing the output of these corticoids the body adapts and maintains effective operations under these stress conditions. Its new level of operating is called the general adaptation syndrome and depends upon the continued activity of the adrenal cortex. We may express this by saying that the feedback cutoff point for ACTH production by the pituitary has been set at a higher level than normally. Under such circumstances the cortex enlarges. If the stresses prove to be excessive, the adaptational mechanism may break down through exhaustion of the adrenal gland, eventuating in the death of the animal.

The activation of the adrenal in the general adaptation syndrome is dependent upon pituitary ACTH, as demonstrated by the failure of adrenal activation in the absence of the pituitary. This stress response is, of course, basically independent of the short-term emergency response mediated by the adrenal medulla and the sympathetic nerves, since the short-term emergencies do not necessarily activate it. Yet it is evident that short-term emergency conditions often lead to long-term stress. Both systems are thus important in behavioral adjustments and must have a common integrating mechanism, which, as we shall see below, lies in the hypothalamus of the brain.

## Hormones Related to Reproduction

The three other tropic hormones of the anterior lobe function in relation to reproduction. One of these, called the follicle-stimulating hormone (FSH), is operative chiefly in development of the germ cells. A second, the luteinizing hormone (LH), strongly influences the secretion of male and female hormone, although some co-operative interaction probably exists between FSH and LH in this respect. Chemical substances having a masculinizing effect are called androgens. A number of steroids secreted by the testes in different animals are the natural androgenic hormones. The most widely used of these is testosterone. Feminizing steroids are called estrogens, estradiol and estrone being the best-known female hormones produced by the ovary.

The third tropin, the luteotropic hormone (LTH)—also called prolactin—has two principal physiological actions. For one thing it stimulates the *corpus luteum* (an organ which develops in the ovary after ovulation) to secrete its own hormone, called progesterone. Progesterone is essential for successful pregnancy because of its action on the uterus and other structures. The second action of luteotropin is that of maintaining the functional activity of the mammary gland, thus permitting successful nursing. In Chapter 4 we will see how these hormones are interrelated in the physiology of reproduction, and in Chapters 5, 6, and 9 we will learn something of their influences on the behaviors related to reproduction.

## Functions of the Thalamic Regions of the Brain

The most anterior region of the neural tube in the embryo gives rise to the forebrain. The midregion of the forebrain consists of a large central mass called the thalamus, a smaller dorsal region containing the pineal gland called the epithalamus, and, finally, a ventral region called the hypothalamus. The floor of the hypothalamus projects downward, forming a funnel-shaped structure called the infundibulum. The pituitary gland is attached to the bottom of the "stem" of this infundibulum. Thus the "master gland" of the endocrine system is intimately associated with the hypothalamus, formed from the floor of the (embryonic) anterior tip of the nervous system (Fig. 2.1.). The interconnections of these two organs are anatomically complex and form the crux of the mechanism of neuroendocrine correlation in behavior. But before we consider this aspect, it would be well to describe the general functions of the thalamic regions of the brain, since these are rarely given their due in elementary descriptions of the nervous system.

Neurologists have sometimes described the thalamus as the "Rome" of the central nervous system (CNS) because all roads through the CNS lead to the thalamus. It receives a vast input of sensory fibers from the brain stem and spinal cord, which bring into this center almost all the information the body receives concerning the environment. The correlating centers in the thalamus, in turn, are connected by two-way relays to the cerebral hemispheres lying on either side. We will have more to say of some of these connections later; here we may note only that the hypothalamus lying below is closely connected to this great information-processing system.

The primary significance of the hypothalamus is that it is the highest outflow (motor) center for control of visceral functions which are mediated chiefly by the autonomic-nervous system and the endocrine organs. By destroying or stimulating particular areas of the hypothalamus these areas have been identified as centers concerned individually with such activities as rage, appetite, wakefulness, sexual development and activity, water metabolism, temperature regulation, etc.

## Relations of the Hypothalamus to the Pituitary

As mentioned above, the pituitary is also under hypothalamic influence, and through it, the sex glands, thyroids, and adrenals are inte-

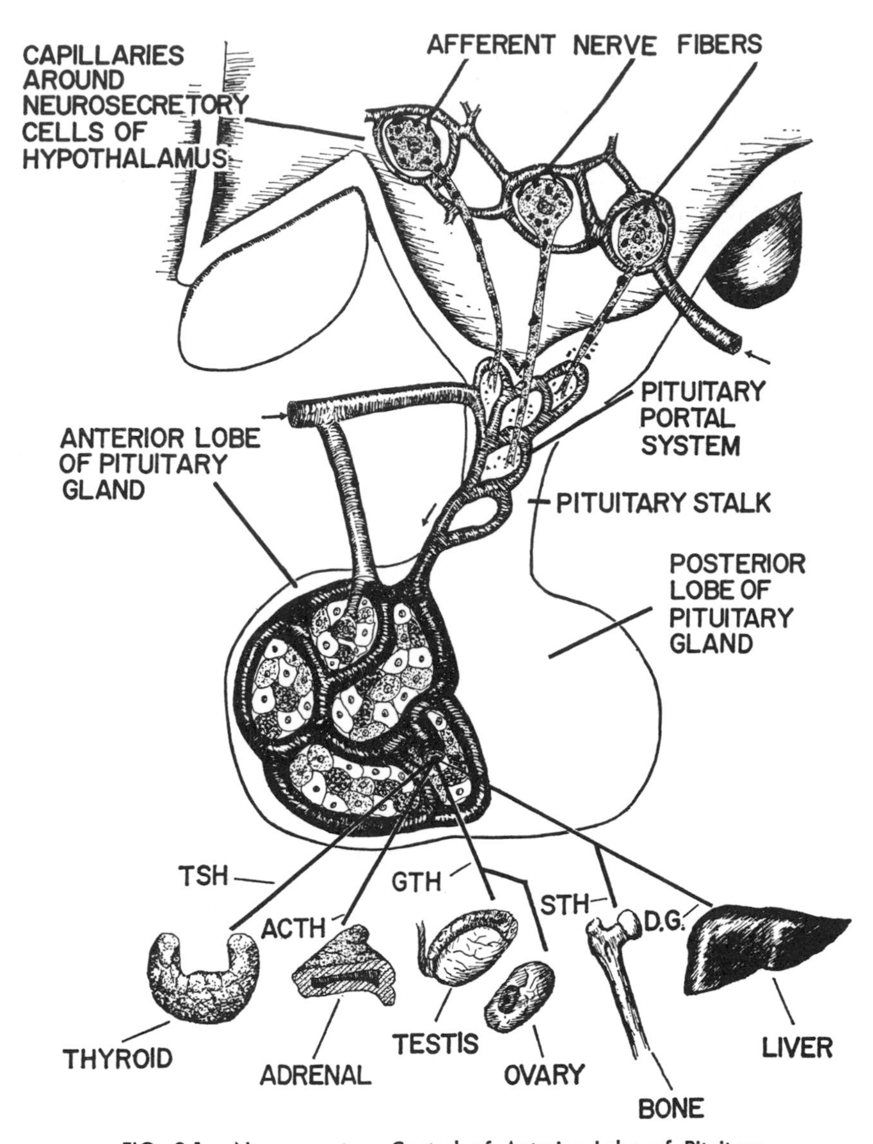


FIG. 2.1. Neurosecretory Control of Anterior Lobe of Pituitary
The floor of the hypothalamus is diagrammed above with three neurosecretory cells in its wall. The axons of these cells are indicated as passing down the pituitary stalk and giving up their neurosecretion to the primary plexus of capillaries of the pituitary portal system, which brings these chemicals into the anterior lobe of the pituitary. Here they influence the cells secreting the various hormones of that gland which pass by way of the blood circulatory system to the organs they control. (Redrawn after F. Netter, The Ciba Collection of Medical Illustrations.)

grated into the overall adaptive responses of the organism. The mechanism of this control was long a mystery, since no nerves seem to reach the anterior pituitary cells from the brain. A series of seemingly unrelated and unexpected discoveries led to an understanding, first,

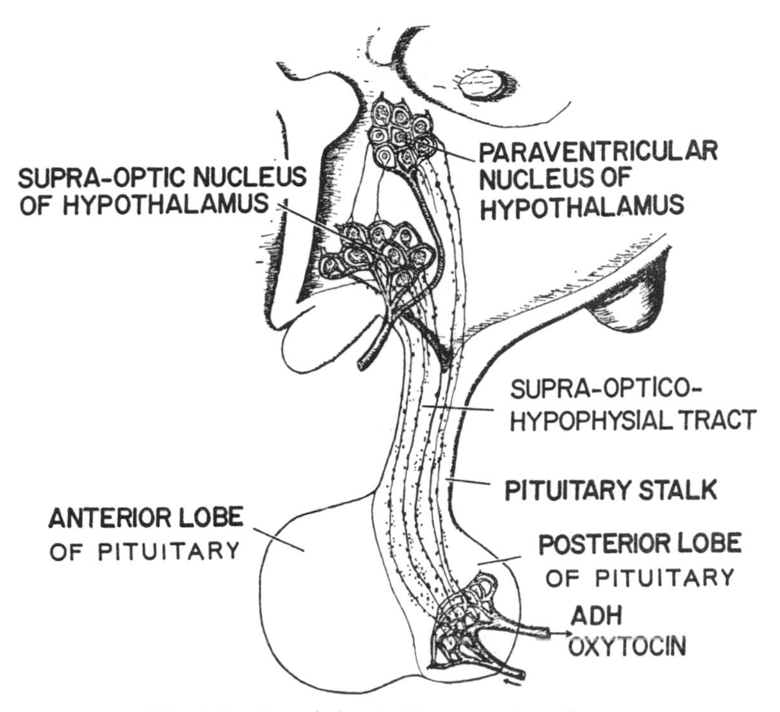


FIG. 2.2. Hypothalamic Neurosecretory System

The floor of the hypothalamus is shown above with the supra-optic and paraventricular groups of neurosecretory cells depicted in its walls. The axons of some of these cells pass down the pituitary stalk in the supra-optico-hypophysial tract to the posterior lobe of the pituitary. The neuro-secretion is stored in the neural part of the posterior lobe (pars nervosa). On stimulation conducted apparently by way of these same nerve fibers, the posterior lobe releases its hormones, oxytocin and the anti-diuretic hormone (ADH), into the general circulation. (Redrawn after F. Netter, The Ciba Collection of Medical Illustrations.)

of the true nature of the neurohypophysis and, then, of the mechanism of hypothalamic regulation of the anterior pituitary.

The basic discovery was the fact that some nerve cells form secretions in the body of the cell and that these secretions are then transported down the length of the nerve fiber to its terminal to be released there, generally into blood vessels. Such neurosecretory cells are found in the nervous systems of many animals. In vertebrates generally, groups of them are found in the anterior part of the hypothalamus in the preoptic nucleus (in mammals, in the supraoptic and paraventricular nuclei). Axons from these cells form the hypothalamico-hypophyseal nerve tract. They pass down the infundibular stalk, ending in the neurohypophysis (Fig. 2.2). This organ thus consists essentially of the swollen ends of the axons in which neurosecretion is stored, ready for release into the capillaries. The neurosecretion contains the two hormones which have long been known to be present in the neurohypophysis. These are oxytocin, which acts on the uterus, and vasopressin, which, besides raising the blood pressure as its name implies, also acts to conserve body water and therefore is also called the anti-diuretic hormone (ADH).

The neurohypophysis releases its ADH in response to water depletion of the body. There is a mechanism in the hypothalamus, probably consisting of cells in the neurosecretory nuclei, which monitors the osmotic concentration of the blood and, when this gets out of the normal range, brings it back again by controlling the release of ADH. Not only is this activity important in the general physiology of the body, but it is of significance in behavior because the conservation of body water is important in stress.

The second apparatus recently discovered is that by which the hypothalamus regulates the tropic activities of the anterior pituitary. It has been found that some of the neurosecretory cells deliver their neurosecretion to the most anterior area of the neurohypophysis, called the median eminence. Here the neurosecretion appears to enter the blood in capillaries which lead into veins that carry it directly into the anterior lobe of the pituitary. These veins thus form a pituitary portal system by which neurosecretion, originating in nerve cells of the hypothalamus, reaches the anterior pituitary to regulate its activity. The nature of the chemicals which are thus transmitted and the pathways through the nervous system by which this regulation is controlled are being actively investigated, and we cannot here attempt to summarize the emerging concepts. However, it is evident that the hypothalamus is the physiological focus of pathways for adaptive responses mediated through the endocrine system, particularly through the sex glands, adrenal cortex, and thyroid (Fig. 2.1).

## Evolution of the Cerebral Hemispheres

As described above, the thalamic regions of the brain are the midline structures developed from the anterior end of the neural plate of the embryo. On either side, hollow bulges form which become the cerebral hemispheres. The ventromedian wall of these bulges, that is, the region adjacent to the thalamus, differentiates into a large mass called the *corpus striatum* or basal ganglia. The rest of the outgrowth forms the cerebral hemispheres proper. In fish these hemispheres are almost entirely concerned with the olfactory sense and have the conventional structure found in other parts of the nervous system, that is, gray matter (cell bodies) on the inside with white matter (nerve fibers) around it. Beginning, however, with amphibians, we find that a layer of gray matter appears outside the white and forms a cortex. The cortex has general correlating functions for all sensitivity received from lower centers, although even in reptiles and birds, where it is much more clearly developed than in amphibians, it is still a secondary correlating center, mostly concerned with olfaction and related visceral functions. In mammals a great development of a new area of cortex appears on the dorsal-lateral surface of the cerebral hemispheres. This new cortex is called the neocortex to distinguish it from the old cortex called archicortex, or paleocortex, of lower forms. The neocortex constitutes the highest correlating center. It is concerned with integrating behavior into environmental changes and is dominant over all lower centers of the nervous system. The neocortex undergoes its greatest development in primates. In man it is larger than all the rest of the brain.

## Relations of the Cortex to the Thalamus

The physiology of the cortex is too complex to be dealt with in any detail here. A few general considerations must be reviewed because they are necessary to appreciate the direction of current research in social behavior.

The sensory input into the thalamus is sent to the neocortex in such a manner that each sensory mode (e.g., sight, hearing, somesthetic sense) is projected to its own area of the cortex. These parts of the cortex are the primary sensory projection areas. Around each is an area concerned with the more complex integration of that sensory input. Between the sensory areas are association areas concerned with still more complex integrations of sensitivity received from different sense modalities. Finally, still other areas, sometimes designated gnostic areas, are believed to be concerned with the highest mental functions.

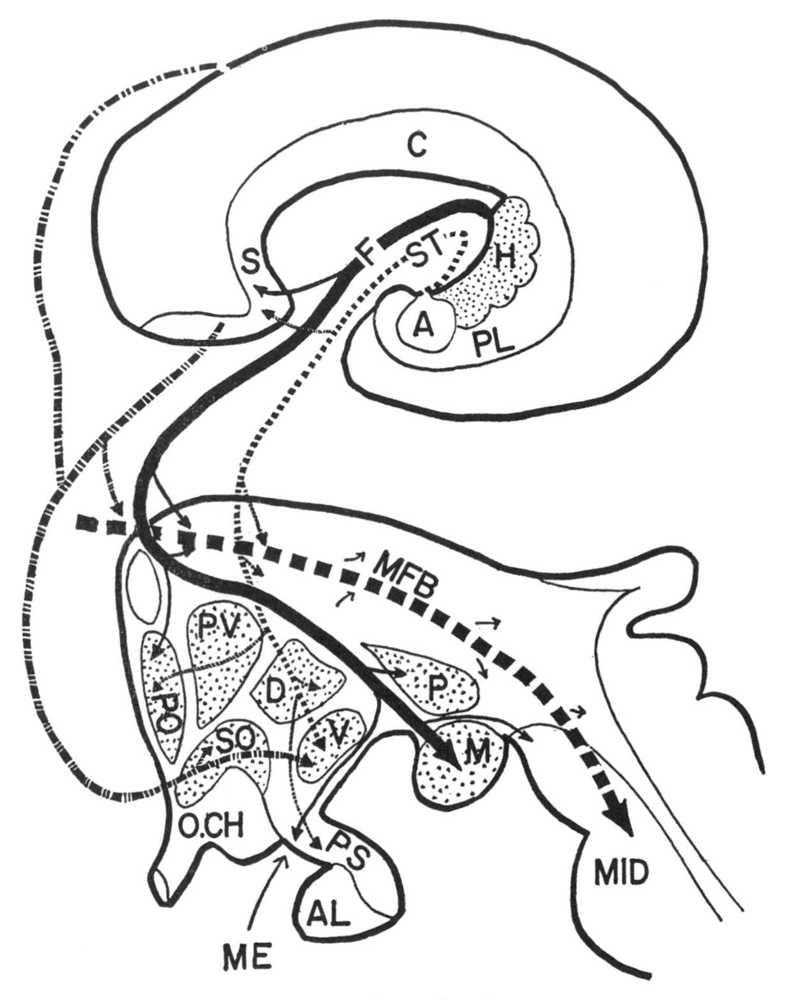


FIG. 2.3. The Limbic System

Those parts of the cerebral cortex of a higher mammal which are derived from the areas of primitive cortex (archicortex and mesocortex) present in lower forms are intimately related to the hypothalamus and to the mid-brain reticulum and through these to the autonomic nervous system. These areas include the cingulate gyrus, the hippocampus, the hippocampal gyrus, the amygdala, and the septal, orbitofrontal, and piriform regions of the cortex. These are inter-connected by many tracts, particularly the fornix, stria terminalis, and the median forebrain bundle. This complex system has been designated the limbic system. In contrast, the main part of the human cortex is the neocortex, which has been elaborated only in higher mammals. The limbic system lies mostly on the median and ventral aspects of the cerebral hemispheres, largely covered over by the great development of the neocortex, and is consequently difficult to show in a diagram.

Studies based on lesions, electrophysiology, and drug action, indicate

The cortex is intimately related to the thalamus. Not only does it receive the sensory projections from the thalamus, but there are extensive two-way connections between the non-sensory nuclei in the thalamus and hypothalamus and the cortex. So intimate is this connection that the two-way interconnection is basic to much cortical functioning because of its provision of indirect connections between different cortical areas. Thus isolation of one region of the cortex from the others by cutting the tissue of the cortex does not necessarily eliminate the functional activity normal to that region, since alternative pathways through the thalamus still connect the region to other parts of the cortex. At any rate, such mental activities as consciousness and emotion probably must be shared in some degree by the thalamus and cortex, with the cortex itself being concerned with the more precise and differentiated aspects of each.

## Relation of the Cortex to the Hypothalamus

Cortical relations to the hypothalamus are of particular interest in social behavior. The closest connections are those between the hypothalamus and the rhinencephalon, which includes those parts of the cerebrum most closely associated with the sense of smell (archicortex). See Figure 2.3. Besides the olfactory lobes and tracts which lie

---

that the limbic system is functionally distinct from the neocortical system. It is concerned in emotional responses. The septal and cingular regions are especially concerned in reproductive behavior and the activities of the endocrine organs. The fronto-temporal region lying more laterally—including the amygdala, the hippocampus, and the hippocampal gyrus—is especially prominent in activities of self-preservation such as feeding and self-defense.

Diagram is based on the work of Gloor and McLean. The upper diagram represents a view of the right cerebral hemisphere seen from the ventro-medial side with the brain stem removed to show the median aspect of the temporal lobe. The lower part of the diagram represents a median section through the brain stem showing the principal nuclei of the hypothalamus in outline.

A—Amygdala
AL—Anterior lobe of pituitary
C—Cingulate gyrus
D—Dorsomedial nucleus
F—Fornix
H—Hippocampus
M—Mammillary body
ME—Median eminence
MFB—Median forebrain bundle
Mid—Midbrain (reticulum)
OCh—Optic chiasma
P—Posterior area of hypothalamus
PL—Piriform lobe
PO—Preoptic area
PS—Pituitary stalk
PV—Paraventricular nucleus
S—Septal area
SO—Supraoptic nucleus
ST—Stria terminalis
V—Ventromedial nucleus

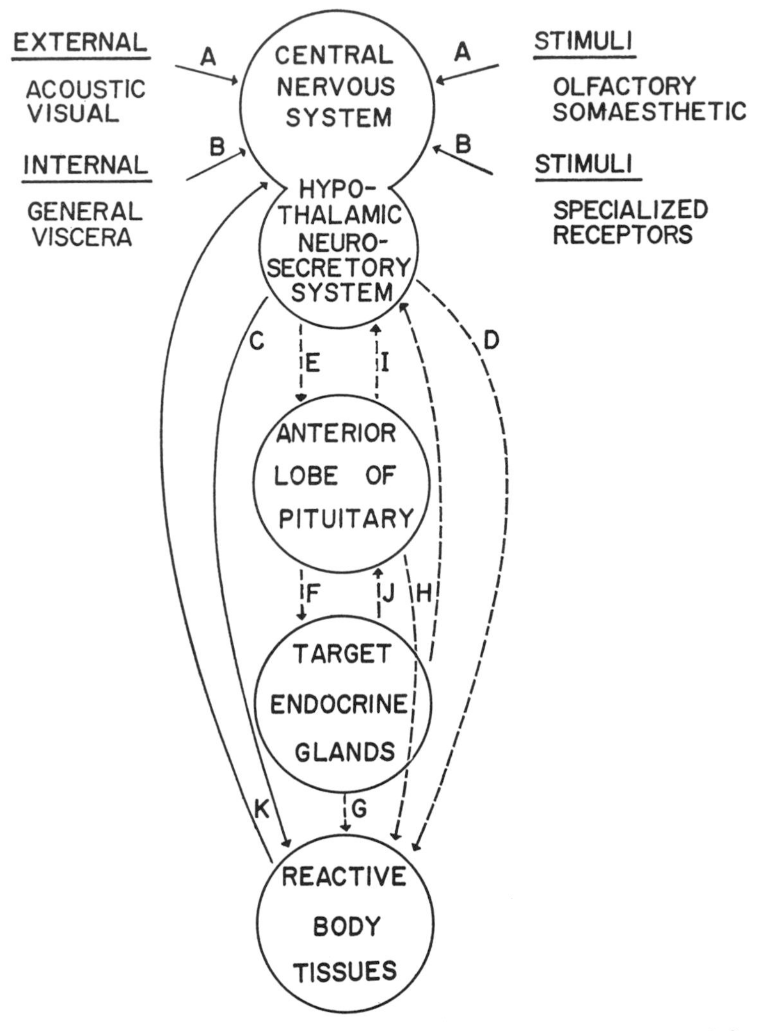


FIG. 2.4. The interrelations of the two principal integrative systems of the body (nervous and endocrine) are summarized in this diagram which is based on the concepts developed in Scharrer and Scharrer (*Neuroendocrinology*). External stimuli, including those emanating from social companions, impinge upon the sense organs and are conveyed to the central nervous system (*A*). Stimuli are also transmitted from the viscera and from specialized receptors (for blood pressure, pH, and osmolarity, etc.) to central nervous centers (*B*). All of these stimuli eventually converge upon the hypothalamic centers governing the autonomic nervous system. Nerves of this system (C) control many body tissues such as sex organs, blood vessels, etc. These same hypothalamic centers influence the neurosecretory neurones in the hypothalamus and control the production and release of their products. Some of these products pass directly into the blood stream by way of the neural lobe of the pituitary and affect some body tissues (*D*). An important component of neurosecretion goes to the anterior lobe of the pituitary

on the ventral surface of the brain, the rhinencephalon includes the cingulate gyrus on the median surface, the hippocampal gyrus and the hippocampal and amygdaloid nuclei on the inside surface of the anterior tip of the temporal lobe. Sometimes this anterior tip of the temporal lobe is referred to as the pyriform area. Another term that has come into use is "limbic lobe," referring to those parts of the rhinencephalon other than the olfactory lobes and tracts themselves. These rhinencephalic structures are the oldest part of the cortex and are present in all tetrapods. The rhinencephalic structures have large tracts connecting them to the hypothalamus, and physiological experimentation has shown that they act as governing centers for integrating hypothalamic activities with the environmental influences that have been received by the sensory input of the thalamo-cortical system (Fig. 2.4). The most marked influence of the limbic system on the hypothalamus seems to be inhibitory. For example, after bilateral lesion of the amygdala in the cat, the animal is reported to show excessive rage reactions. It is important to understand that the cortex, as the highest integrating center, governs, not only intellectual functions such as learning, but also the visceral activities and emotional life of the animal (Fig. 2.5). It is probably through the connections of the limbic system to the hypothalamus that many of the psychosomatic effects of emotional experiences are mediated. In Chapters 3, 5, and 7, we shall see that such effects play a significant role in the social life of animals, including man.

## Functions of the Reticular Formation

Another area of the brain that bears important functional relations to the hypothalamus and the limbic system is the so-called reticular for-

---

(*E*) by way of the median eminence and the portal blood vessels. From the anterior lobe of the pituitary it influences the cells secreting the tropic hormones (*F*) which control the rate of secretion of the hormones of the target endocrine glands, such as the thyroid, adrenal cortex, and sex glands. Hormones of these target glands act upon the body tissues (*G*). Other anterior pituitary hormones (growth hormone and prolactin) act directly on body tissues (*H*). Feedback of various hormones influences the activity of the neurosecretory system (*I*), thereby making the hypothalamic pituitary system a self-regulatory one. Similarly, feedback by the hormones of the target endocrine glands (*J*) helps control anterior lobe functions. By way of sensory nerves (*K*) the body tissues involved in behavior also feedback into the central nervous system. Simplified as this schema is, it shows clearly that an influence impinging on any point of the system may disturb its equilbrium in many different ways.

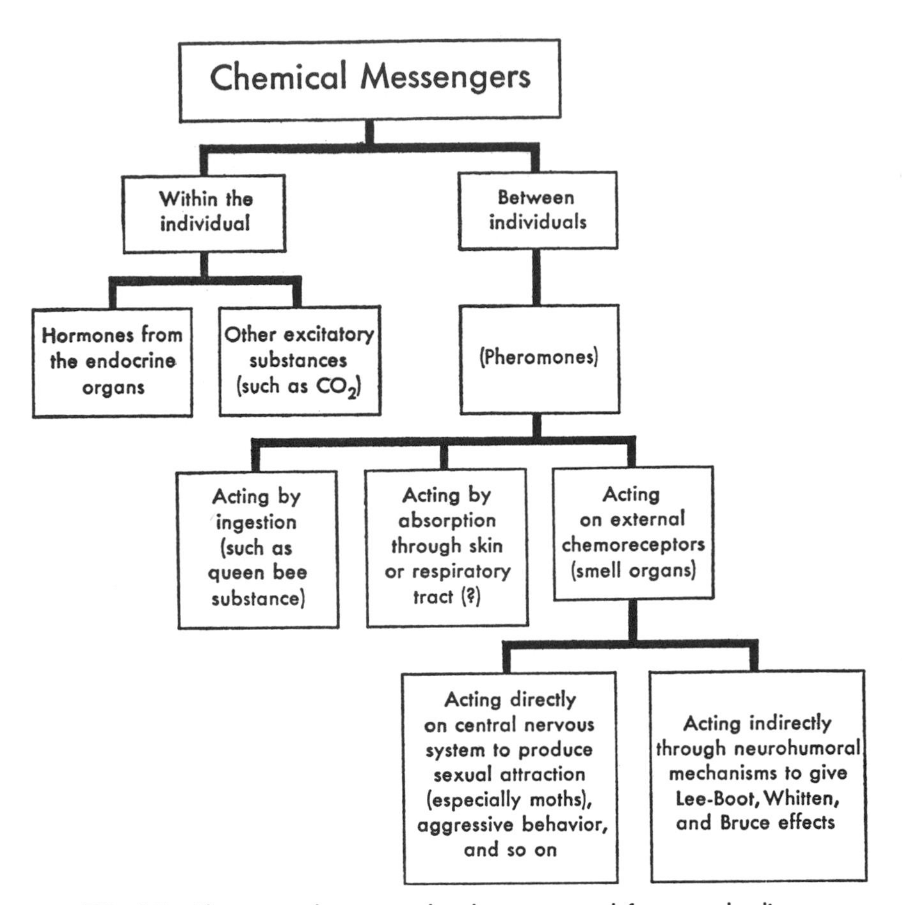


FIG. 2.5. The name pheromones has been proposed for a newly discovered class of substances which are produced by some individuals in a population and transmitted to others whose physiological processes they influence. In a sense, then, such substances act like external hormones. Several examples of such are known in social insects, where they serve to regulate production of members of different castes, e.g., soldiers among termites or queens among bees. Among mammals odoriferous substances produced by one sex are known to have profound effects upon the physiology of the opposite sex. For example, in mice, the smell of strange males has been shown to inhibit pregnancy in recently mated females (Bruce effect). On the other hand, females when housed together influence each other in the direction of prolonging their estrus cycles (Lee-Boot effect). The introduction of males to previously isolated females also strongly influences their estrus cycles (Whitten effect). Since these effects are produced even when contact between the mice is prevented by inclosing them in separate wire frames but are prevented when the olfactory sense is destroyed, the effects are believed to be transmitted by odoriferous substances. Since reproduction in mice is regulated by the pituitary the paths of action of these substances must be from sense organ, through brain, through the neuroendocrine mechanism of the hypothalamus to the pituitary. (After Parkes and Bruce, *Science*, 134 [1961].)

mation. This consists of central areas in the spinal cord, brain stem, and thalamus which are morphologically primitive in their structure. In them the nerve cells and their fibers form a diffuse network rather than being segregated into nuclei and fiber tracts as they are in more differentiated parts of the central nervous system. This system receives collaterals from the sensory inflow into the central nervous system at various levels of the body and transmits these impulses through a series of short relays to higher centers. It similarly receives motor output from higher centers and transmits it by short steps to the motor neurones of lower centers. Transmission through the reticular formation is thus slow and diffuse, allowing for many synaptic interconnections along the way. The reticular formation in the midbrain has been found to play an important role in regulating the general level of cortical activity. After lesions in this part, which is sometimes designated the reticular-activating system, the animals tend to sleep excessively. Even minor interference with this system hinders the maintainance of attention by the animal. Apparently a steady barrage of impulses from this system helps maintain the level of cortical activity necessary for consciousness and attention.

Some of the motor functions of the reticular formation are clearly defined. For example, one area in the medulla serves as a respiratory center, integrating and controlling the activities that maintain appropriate breathing rates. In addition to these definite integrating centers, the reticular formation throughout the nervous system acts as one of the principal pathways of impulse-transmission from the hypothalamus and limbic thalamo-cortical systems to the autonomic system. It is thus strategically located in the control of visceral functions of the body (Fig. 2.3).

## REFERENCES

Recent works giving up-to-date information on the materials covered by this chapter are:

Field, J., Magoun, H. W., and Hall, V. 1960. *Neurophysiology.* 3 vols. (*Handbook of Physiology,* Section I.) Washington, D.C.: American Physiological Society.

Gorbman, A., and Bern, H. 1962. *A Textbook of Comparative Endocrinology.* New York: John Wiley & Sons.

House, E., and Pansky, B. 1960. *A Functional Approach to Neuroanatomy.* New York: McGraw-Hill Book Co.

Jasper, H. H. (ed.). 1958. *The Reticular Formation of the Brain.* Boston: Little Brown & Co.

Netter, F. 1958. *The Nervous System.* Summit, N.J.: The Ciba Co.

PIELE, T. 1961. *The Neuroanatomic Basis for Clinical Neurology.* 2d ed.; New York: McGraw-Hill Book Co.

SCHARRER, E., and SCHARRER, B. 1963. *Neuroendocrinology.* New York: Columbia University Press.

TURNER, C. D. 1960. *General Endocrinology.* 3d ed.; Philadelphia: W. B. Saunders Co.

YOUNG, W. C. (ed.). 1961. *Sex and Internal Secretions.* 3d ed.; Baltimore: Williams & Wilkins Co.

DAVID E. DAVIS

# 3

# The Physiological Analysis of Aggressive Behavior

## Introduction

Aggressive actions are among the most prominent social activities of animals, including man. Such actions often appear antisocial, but the fighting, bluffing, and threatening may serve to promote, with some individual suffering, survival of the species. In this chapter aggressive behavior will be operationally defined as action involving two or more animals that threaten, bluff, or attack each other. The action may result in physical combat, but, more frequently, one individual flees before combat starts or at least before injury occurs. Learning plays a prominent role, since an individual may learn through experience that a particular individual or a type of individual can inflict damage. When such learning has occurred, one individual will retreat or flee at the sight of another. For example, young woodchucks retreat or watch carefully even when an adult is as far as from seventy-five to a hundred feet away. Although no contact occurs, social behavior is emphatically present.

Aggressive behavior is widespread throughout the animal kingdom. Among the invertebrates the behavior of many insects and molluscs which attack members of the same species has, unfortunately, been described only in anecdotes. Comparisons with vertebrate behavior are difficult, since the nervous and endocrine systems are fundamentally different. Among the vertebrates, aggressive behavior is found in species from fish to man. Relatively few species of fish have been

David E. Davis is professor of zoology at Pennsylvania State University.

studied, and these represent the most advanced forms (teleosts). Species of the classes which include sharks, hagfishes, lampreys, lungfish, and other primitive groups need examination. Among amphibians and reptiles, aggressive behavior has been observed but little physiologic work done. As might be expected, birds and mammals have been frequently studied from both the behavioral and physiologic viewpoints.

The reciprocal effect of behavior on physiologic processes has recently attracted attention and research. This chapter considers only those features of aggressive behavior now known to affect physiologic processes, the study of which promises to produce great advances in our knowledge of physiology. Thus, many psychologic and psychiatric aspects (McNeil, 1959) are completely excluded. Also, many homeostatic processes are ignored.

The term "aggressive behavior" usually excludes predaceous behavior or defensive actions against a predator. The flight of a rabbit from a dog or the alarm cries of nesting birds at a cat are considered defensive, rather than aggressive, social behavior. The attack on a field mouse by a red-tailed hawk also is not considered to be aggressive behavior. It is true that these actions include attack, threat, and defense and that many muscular actions in a fight are the same as those occurring in social behavior. The distinction between predaceous and aggressive behavior rests, however, on two main features. First, predatory attacks (with some rare exceptions) are not upon the same species, i.e., they are interspecific. Second, the actual behavior leading up to the predatory attack is different from that shown in intraspecific aggression. When a dog is ready to attack another dog to settle problems of social behavior, he uses motions of the tail, head, etc., different from those he uses preparing to attack a cat. A complicating feature of this distinction is that the physiologic repercussions (to be described in detail later) appear to be very similar, if not identical, in the two types of behavior. However, little has been done on the physiologic consequences of predaceous or defensive actions, in contrast to aggressive actions, principally because physiologists have not recognized the differences between these two types of behavior (Fig. 3.3).

Aggressive behavior is also clearly distinct from sexual behavior. Proof of this statement will not be documented in detail here, since it is sufficient to point out certain aspects of the existing evidence. No correlation exists between the amounts of aggressive and of sexual behavior shown by an individual (if social rank is controlled) in various captive birds and mammals. Indeed, the most aggressive hens gen-

FIG. 3.1. Start of aggressive behavior between cows

erally lay fewer eggs than their less aggressive pen mates. Observations of wild species frequently show aggressive behavior appearing without any sexual features in the fall.

The actual patterns of aggressive behavior differ from species to species. Detailed descriptions are readily available in current literature and may be summarized briefly. Fish generally attack by butting and nipping. Lizards threaten by flashing the dewlap (skin under the throat). Birds attack with bill, wings, or feet, or threaten and bluff by song and by displays involving exposure of bright or striking colors or patterns. Mammals use a variety of patterns to intimidate another member of the same species. For example, cows and sheep butt heads, mice vibrate their tails and hunch their backs, and primates threaten vocally (Figs. 3.1, 3.2, and 3.4). At the present time there is great need for detailed descriptions (perhaps from moving pictures) of behavior patterns in aggressive social actions. One more point requires emphasis. The aggressive action may be very subtle, and its detection may depend on remarkable powers of observation and of induction. For example, cows, apparently placidly grazing, demonstrate their aggression by trivial changes of direction. The subordinate veers away gradually from the dominant cow, even though the latter may be at a distance of ten or twenty feet. The detection and interpretation of these subtle behavior patterns is a challenging task.

FIG. 3.2. Butting used by cows to decide rank

**FIG. 3.3. Defense behavior of lemmings against a predator. The threat posture is shown, and the leap into the air is shown, *center* and *right*. (Redrawn from Arvola, Ilmen, and Koponen, *Arch. Soc. Zool. Bot. Fenn. Vanamo*, 17 [1962].)**

The aggression involved in social behavior evolves out of two fundamental patterns. Some species divide an area into territories and others arrange themselves into a social rank. These patterns may be at the poles of one continuous phenomenon (Davis, 1958), but for the present we can consider them distinct. They are discussed in more detail in Chapter 1.

The major function of aggressive behavior is to determine and maintain rank or territory. In every species, regardless of its type of social organization, the aggressive individuals assert their position and maintain it by certain behaviors, some of which were mentioned above. In this action, many physiologic processes participate, some stimulating aggressive behavior and others responding to it. The next two sections of this chapter will consider these processes in some detail.

An unsolved problem or set of problems is the extent to which aggressive behavior is learned. Presumably, many details are learned or, at least, perfected through experience. Other aspects, especially the disposition to aggressiveness, are apparently innate. Experimental evidence is meager, but some items are relevant. Most of the work has been done with domesticated mice, but the little work done with some other species confirms the results. It is commonly known that male mice isolated at weaning will fight when put together at three or four months of age. Since it appears unlikely that the fighting is learned during suckling, it is inferred that the disposition to fight is innate. Early social experience nevertheless greatly affects subsequent aggressive behavior (King, 1957). As Scott (1958) has pointed out, the learning of fighting follows psychologic principles. When male mice raised in isolation are placed together, most individuals will start to fight. Within a few hours or days a social rank is organized, and the mice have learned their places. Fighting then becomes rare, since the subordinate mice no longer fight. If a dominant mouse is put with other mice, he will again fight and usually win. In contrast, subordinate individuals will rarely assert themselves. However, by special techniques consisting of artificially producing a "win" by the subordinate mouse, it is possible to teach the formerly submissive mouse to fight and to win. With another type of experiment, it is possible to train some mice, while still relatively young, to win and others to lose. These individuals then will maintain their statuses unless another

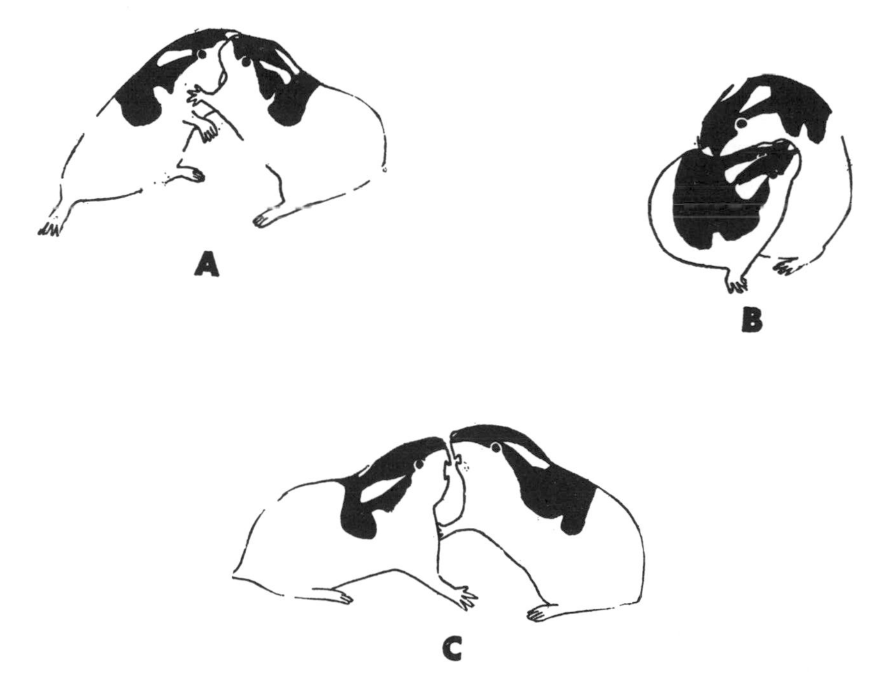


FIG. 3.4. Fighting between male lemmings. (*A*) Initial phase: The animals box with each other. Each bites the opponent's snout. (*B*) Second phase: The animals roll into a tight ball and bite each other in the fore part of the body. (C) The threat posture is adopted by the inferior animal (*left*). This makes the opponent retreat. (Redrawn from Alvola, Ilmen, and Kaponen, *Arch. Soc. Zool. Bot. Fenn. Vanamo,* 17 [1962].)

teaching program reverses it. It is clear from this type of research that learning plays an important role in modifying the physiologic basis of aggressive behavior.

Recent work suggests that the object of aggression is learned but the actual motions are innate. Blauvelt (1960) raised goats and sheep under three situations: a goat with a goat, a sheep with a sheep, and a goat with a sheep. In each combination the animals, when tested with a strange individual of either species, attacked the kind with which it had been raised. For example, when a sheep which had been raised with a goat was put with a sheep, no fight occurred since the sheep did not attack the other sheep. In contrast, when the sheep was put with a goat, it attacked. However, it fought by butting in the manner of a sheep rather than by jumping in the manner of a goat. No effective fight occurred because the sheep tried to butt the goat and the goat tried to jump on the sheep. No contact was maintained and the attacking soon ceased. These few experiments should be amplified and extended to many species to clarify the problem of the role of learning in aggressive behavior.

The measurement of aggressive behavior may present problems. In many cases, a simple statement that such behavior did or did not occur is adequate. Thus, one can say that in one treatment 70 per cent of the animals fought, but in another, only 30 per cent fought. More sensitive measures, perhaps, are "latency," which is the time till aggression starts and "duration," which is the length of time aggression lasts. Other more intricate measures may be used. All procedures should be organized so that statistical tests of significance can be made.

## Physiologic Factors in Aggressive Behavior

A major purpose of this chapter is to place in perspective the physiologic factors in social behavior. To understand the causation of behavior, it is necessary to consider the hormones as "agents" that carry a message concerning a state or condition. Hormones are a means of communication supplementing the nervous system but sometimes acting in different ways (see Chap. 2). Hormone action is slow because the chemical must travel in adequate quantities from the gland to the site of action. When the hormone is present at the proper structure, however, action may be almost instantaneous. The well-known rapid action of adrenalin is an example. Even testosterone, when placed at a suitable place in the hypothalamus, promptly elicits sexual behavior (Fisher, 1956).

In many vertebrates the glands change in size seasonally so that many weeks are necessary in the spring to produce the hormones. The behavior may develop gradually and increase in frequency. As an example, cardinals begin to sing in early January, but their song is haphazard and weak, and their defense is mild. As the season progresses, their song becomes vigorous and their defense of territory becomes active. The appropriate hormones are now present in adequate quantities.

In addition to the hormones, an appropriate stimulus situation must always be present. Thus, a bird may defend its territory vigorously but feed peacefully with other birds in some common area away from the territory.

The "agent" cannot deliver its message unless suitable conditions are present. Thus, testosterone, when injected into a capon, can cause no behavioral expression unless another bird is present and reacts in certain stereotypic ways. Furthermore, the environment must be suitable (familiar; contain certain items). For example, some birds will not demonstrate aggressiveness unless a particular type of perch is available. Naturally, animals in nature are usually found in areas suitable simply because the animals chose them. Work in laboratories, however, often fails to mimic natural conditions, and hence the "agents" cannot deliver the messages.

The role of hormones in aggressive behavior has been recognized for a century. The frequent observation that roosters when castrated soon cease fighting is amply confirmed by detailed studies. Many species have been experimentally tested, either by injection of an androgen such as testosterone or by castration. Some of the species are swordtail fish, anolis lizard, painted turtle, domestic fowl, night herons, doves, rats, mice and, last but not least, boys. Other cases are mentioned by Scott and Fredericson (1951). In may cases female or sexually immature individuals respond to androgens as characteristically as do males. In addition to the aggressive behavior, several other masculine features, such as crowing by a rooster, appear. This almost universal reaction of animals to androgens unfortunately has permitted the assumption that aggressive behavior always depends upon androgens. An exception will be mentioned below.

An important problem that confuses the interpretation of results of injections with hormones is the production of anesthesia by high doses of steroids. Testosterone given in doses of 2.0 mg. per chick per day will cause many individuals to become so sleepy they may neither feed properly nor show any aggressive behavior. This unexpected action of testosterone may explain some of the apparently negative results obtained, for example, by Bevan *et al.* (1957). To determine the influence of hormones on aggressive behavior, one must conduct enough experiments to determine dose-responses for several different doses of the hormone. Chicks given testosterone daily in the following doses: 0, 0.25, 0.50, 1.00 and 2.00 mg. showed the following weights of the comb after twenty-five days (average of four chicks): 380, 581, 704, and 643 mg., respectively. Obviously the high dose was detrimental. Those birds also had smaller testes, weighed less, and were very inactive. These data are cited as an example of some of the pitfalls in behavioral work using hormones. The current status of the work on androgens indicates that aggressive behavior is greatly accentuated in many species when the hormone is injected or is naturally present.

Females fight much less vigorously than do males but nevertheless show aggressive behavior that can often be accentuated by injection of testosterone. Presumably the endocrine glands in females produce enough androgens to stimulate aggressive actions. In mammals some special androgens apparently arise in the adrenal glands, and in birds androgens come from the ovary as well.

The hormones (estrogens) normally associated with female characters apparently stimulate aggressive behavior under some circumstances. For example, Birch and Clark (1946) and Kislak and Beach (1955) found evidence that estrogens accentuated aggressiveness in chimpanzees and hamsters. The female hamster, however, is remarkable in its inconsistent behavior.

Since the production of male and female hormones depends upon the activities of several organs, it is now necessary to consider these individually in order to explain the source of aggressive behavior. It is clear that the gonads as well as both parts of the adrenal are controlled by an area in the brain called the hypothalamus through its control of the pituitary (discussed in Chap. 2). The hypothalamus influences aggressive behavior through many channels. Through its control of the pituitary it acts upon the gonads and the adrenal cortex. The gonads, in turn, produce testosterone, which promotes aggression. The adrenal cortex produces corticoids which act upon the organs involved in such behavior. Acting through the sympathetic nerves to the adrenal medulla, the hypothalamus also controls the release of adrenalin, which is concerned in the physiologic processes active during aggressive behavior.

Until recently the hormones of the pituitary were not known to affect aggressive behavior directly. However numerous observations of social behavior in nature suggested that testosterone could not be responsible for the level of aggressiveness. Davis (1957) showed, however, that testosterone did not affect the rank of starlings in the group hierarchy. Subsequently, Mathewson (1961) showed that luteinizing hormones (LH) increased aggressiveness and caused birds to reverse rank (Fig. 3.5).

Some details of this work illustrate how research in aggressive behavior may be conducted. Several types of experiments were done, but one type demonstrates the relationships particularly well. Two birds which had been kept in different cages were put together in a large cage till one clearly showed dominance. Then the subordinate bird was given 0.5 mg. of LH. In six out of eight pairs studied, the subordinate bird assumed a dominant rank. Birds injected with saline did not change rank.

This recent discovery opens a new area in the study of social behavior and illustrates several points of general biological interest. First, it reveals another case of the utilization by a species of a hormone to influence a process (in this case, behavior). Animals put hormones to use in a variety of ways. Seemingly, chemicals serving only a few functions were applied during evolution to additional functions. At the moment we cannot say whether LH, testosterone, or some other hormone was the first to be associated with aggressive behavior. A speculation may be made, and perhaps it will be tested some day. Since one function of LH is to stimulate production of testosterone through the Leydig cells of the testis, it would seem that LH is more basic and the connection of testosterone with aggression is secondary. From another viewpoint, the fact that testosterone is important in specialized birds (domestic fowl) and LH, in generalized birds (starlings) suggests that LH is more basic. Until more information is available, further speculation should be restrained.

From the ecological viewpoint, the separation of aggressive behavior and sexual behavior may be advantageous to some species and may have occurred in the evolution of some song birds. Although evidence is not yet abundant, it is clear that in some species, at least, the male selects and defends a territory and also settles social rank problems in the fall. Androgens cannot be responsible because the as

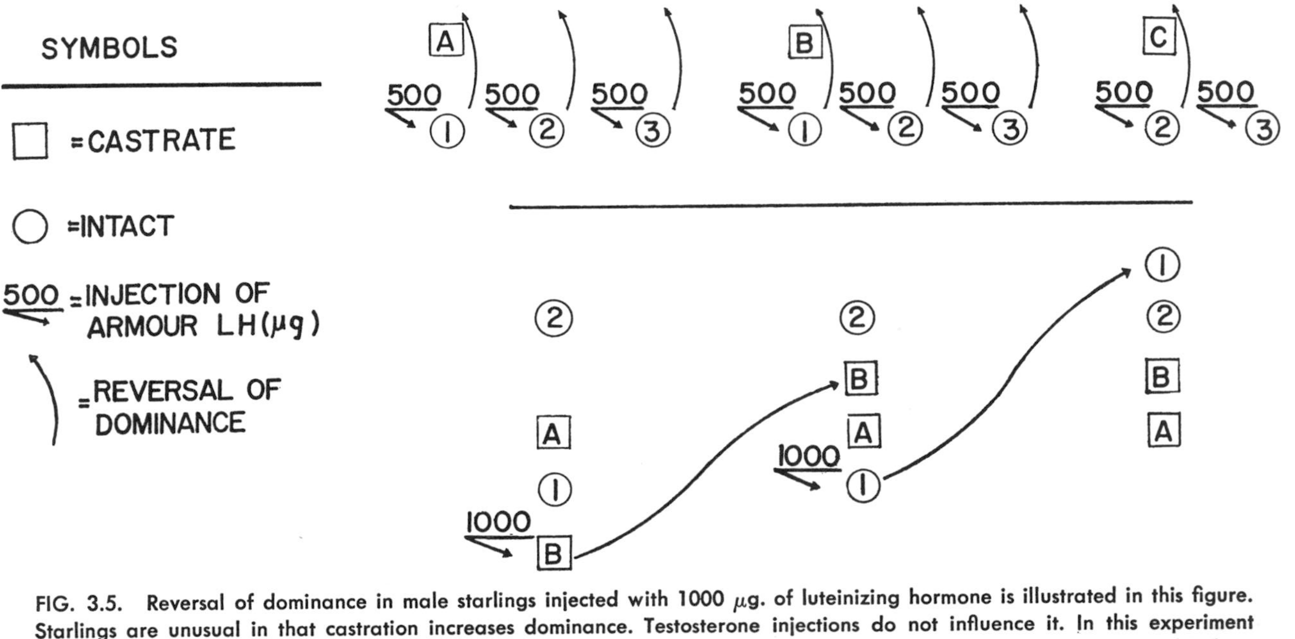


FIG. 3.5. Reversal of dominance in male starlings injected with 1000 $\mu$g. of luteinizing hormone is illustrated in this figure. Starlings are unusual in that castration increases dominance. Testosterone injections do not influence it. In this experiment pituitary hormone (LH) was shown to increase dominance. The bird which controlled the only perch available was the dominant. In the upper figure, A, which originally dominated 1, 2, and 3, was displaced after the subordinate bird was given an injection of 500 $\mu$g LH. B was similarly displaced. Only with C did one injected bird (3) fail to achieve dominance. Dominance shift of the lowest bird in A's hierarchy of four after injection of 1000 $\mu$g LH is shown in the lower figure. (After Mathewson, *Science,* 134 [1961].)

yet undeveloped gonads are tiny. When the breeding season arrives, these males have an advantage over others that have not settled. In spring perhaps another set of hormones (androgens) is responsible for the sexual behavior. A separation of function of the two kinds of hormones allows greater efficiency, since each type of hormone has a separate job. The induction of song by testosterone occurs in several species, however. Song usually has at least two functions: defense of territory and advertisement to the female. Perhaps in some species the latter is the major function, or, possibly, defense is an acquired function.

Several hitherto unexplained behaviors may be now clarified by experimental or field study. For example, many birds (e.g., blackbirds) defend a territory in the fall and winter when the gonads are very small (Snow, 1961). The usual explanations have been that minimal amounts of androgens are derived either from the tiny gonads or from the adrenals. It may be that the LH produced by the pituitary is responsible. It has long been known (Greeley and Meyer, 1953) that gonadotropins (LH and FSH) are produced in large amounts in the fall. Another example of unexplained behavior is the aggressive behavior of female mammals immediately after copulation. It is well known (Sawyer, 1959) that LH appears in large amounts at ovulation in mammals and that it stimulates ovulation in birds. Aggressive behavior occurring at this time might result from LH. A further example may be the often-reported reversal of dominance for pairs of birds at the time of laying.

The preceding paragraphs are admittedly speculative but are included for two reasons. First, it is noteworthy that the observation of wild animals (starlings) in the field of social behavior led to a discovery in endocrinology. Second, the problem suggests many studies (either observational or experimental) feasible in modest laboratories because they can be done with a minimum of space and equipment.

The hormones mentioned thus far are produced in the gonads or pituitaries and are concerned with reproductive activities. Several additional hormones (progesterone, relaxin, etc.) are associated directly with reproduction, but investigations have thus far not demonstrated that they have any influence on aggressive behavior. Because there are no suggestions that such hormones influence aggression, only a few studies have been made. Other organs, such as the adrenal, thyroid, and pancreas, produce various hormones. But only certain hormones from the adrenal are important in the study of aggressiveness.

As discussed in Chapter 2, the adrenal gland is a complicated structure composed of two very different parts, the cortex and the medulla. The cortex encompasses the medulla and consists of several zones (Chap. 2). Some areas produce various hormones called glucocorticoids, which affect carbohydrate metabolism, certain reproductive processes, and inflammation. Other areas produce mineralocorticoids and affect the mineral balance. Also, some androgens appear. The direct measurement of corticoids and androgens is very difficult. Usually the best procedure is to determine the amounts of breakdown products in the urine. Since this measure requires careful interpretation, the results are often dubious.

The medulla may be considered part of the autonomic nervous system, since it develops from the same embryonic source and is homologous with a sympathetic ganglion. In keeping with this origin the medulla produces hormones—adrenalin (epinephrine) and noradrenalin (norepinephrine)—that are collectively called catechol amines. Since a number of detailed reviews (Elmadjian *et al.*, 1958; Christian, 1963) are available, only a summary of pertinent functions will be given here. In general the effects of these two medullary hormones mimic those of the sympathetic nervous system. Both hormones act on the circulatory system to constrict the visceral vessels, but noradrenalin constricts the vessels in muscles while adrenalin dilates skeletal muscles. Adrenalin causes a greater rise in blood sugar than does noradrenalin. Both hormones have various effects on smooth muscle, the spleen, and the bladder, but an effect considered important for current purposes is the mobilization of sugar from the liver. Adrenalin is more effective than is noradrenalin, but the end result is to move carbohydrate from the liver to the muscles.

The sympathetic nervous system responds promptly in a difficult situation, whether it be environmental (cold, poison, etc.) or behavioral (sexual, aggressive). The various effects of the hormones prepare the animal for action in the classic "fight or flight" picture. Recent studies (Elmadjian *et al.*, 1958) suggest that aggressive or active situations tend to stimulate production of noradrenalin whereas tension produces more adrenalin. Thus, professional hockey players had a sevenfold rise in noradrenalin and a threefold rise in adrenalin during a game. Two men who sat on the bench had a trivial change in noradrenalin and a doubling of adrenalin. Several other studies with prize fighters confirmed these results, as did studies on psychotic patients, although the data here are meager at this time.

The pathways in the brain that conduct stimuli to secrete adrenalin are not known. Experimental work shows that the hypothalamus is involved, but the mechanism remains to be convincingly demonstrated. Certainly hormones (vasopressin) from the posterior pituitary can release adrenalin, but many details need clarification. Innumerable studies show that an increase in catechol amines follows psychologic stresses, presumably acting through the limbic region (MacLean *et al.*, 1960). Further discussion of this problem will be reserved for the section on emotions.

While the action of the hormones from the medulla produces this vast array of physiologic responses, it is clear that these are merely symptoms of a situation leading to aggressive behavior rather than causes of it. Adrenalin permits an animal to carry out aggression rather than causes it to be aggressive. For example, persons receiving adrenalin demonstrate the physiologic reactions but do not feel mad or become aggressive. The physiologic responses of the adrenal medulla are merely concomitants of the phenomenon called emotion, which will be discussed later.

## Physiologic Consequences of Aggressive Behavior

The knowledge of the physiologic repercussions of aggressive behavior has a long and complex history. Many miscellaneous bits of information eventually resulted in a hypothesis to be tested. As so often occurs in science, the real advance came from an unexpected direction. The history in brief was as follows: In studies of animal populations, various persons noted that at high densities the reproductive rate declined. Introduction of strange individuals into a population also resulted in a decline, presumably because of social turmoil (Davis, 1949). A little later Christian (1950) proposed that the crash of populations in the rodents of Arctic regions resulted from exhaustion of the adrenals due to the added stress of reproduction and other factors.

Subsequent research showed that social behavior caused changes in the adrenal cortex, resulting in a decline in reproduction. The evidence (given in detail by Christian, 1963) is as follows: When certain mammals are brought together in groups, fighting begins and the adrenal cortex enlarges, reaching a maximum in about ten days. The relationship between population density and size of adrenal cortex has been demonstrated in mice, rats, woodchucks, voles, monkeys, and chickens and presumably occurs in other species that form social ranks. When mice were put together in cages containing 1, 4, 6, 8, 16, or 32 males (same age and weight), the weight of the adrenals increased from 4.92 mg. to 5.33 mg. at 16 animals per cage but dropped to 4.95 mg. at 32 animals per cage. Numerous other experiments confirm that grouping results in an increase of adrenal size, until, at high density, exhaustion occurs (Fig. 3.6). Another type of experiment consists of permitting a few pairs of animals (usually mice) to reproduce in a large cage. In about six months the population reaches a high level. Under these circumstances the size of the adrenals increases about 25 per cent over the adrenal size of controls (mice in isolation). At the half-way point of population increase, the adrenals

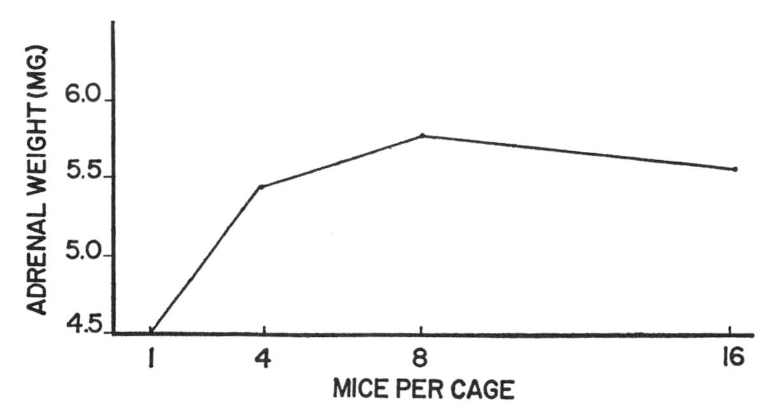


FIG. 3.6. Increase of adrenal weight with increase in number of mice in a cage

weigh about 15 per cent more than those of controls. Similar results occur in chickens (Flickinger, 1961) and, presumably, in other birds. The adrenals were larger in chickens kept in groups of six than in those kept in pairs.

The above results were obtained with caged animals and might be only artifacts of domestication. In natural populations of rats, however, the adrenals in dense populations were about 20 per cent larger than the adrenals of rats in sparse populations. Furthermore, the experimental reduction of natural populations of rats had the predicted result, namely, a reduction of 26 per cent in adrenal weight. Other species, such as meadow voles, show comparable results. Some reports about animals in nature appear to contradict the above statements. Unfortunately, conditions in nature are difficult to control, and densities difficult to compare. For example, woodchucks at a time of high numbers in the fall actually have a lower density as measured by aggressive behavior than they have in the spring. Under these circumstances, apparently negative results must be treated with caution.

The above measurements ignore the social organization of the group. Experiments designed to compare adrenal weight with rank showed that the adrenals of the dominant mice were the same size as those of controls but the adrenals of subordinate mice were enlarged (Fig. 3.7). This relation also occurs in roosters. Thus, an im-

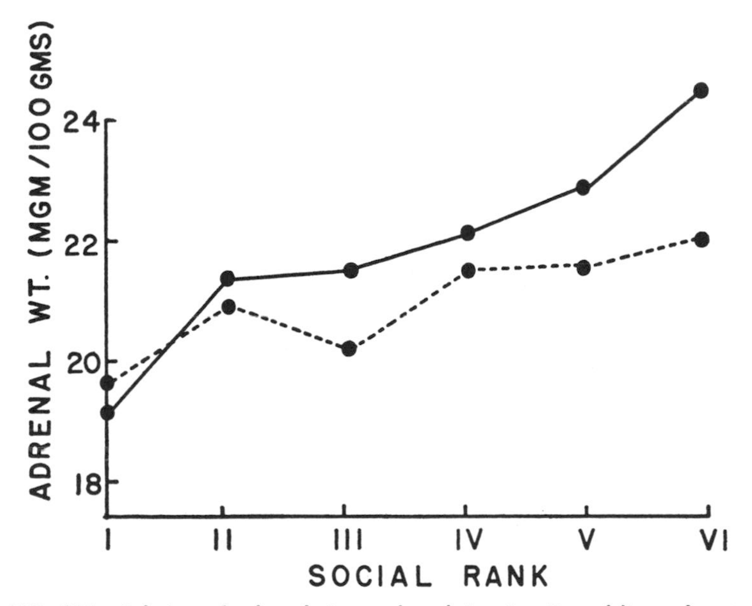


FIG. 3.7. Relation of adrenal size and rank in mice. Dotted line refers to weight at start and solid line refers to weight at end of ten days.

portant consequence of aggressive behavior is the physiologic advantage gained from winning, since the adrenals of winners are normal in size rather than hypertrophied and thus have normal function—we assume.

A number of indirect measures show that an enlarged adrenal secretes increased amounts of hormones. One specific measure is the degree of involution of the thymus gland, which is a lymph organ in the throat region. Injection of corticoids involutes the thymus, as does grouping animals under conditions conducive to aggression (Fig. 3.8). Another indirect measure of adrenal activity is the condition of the gonads. The testes of male animals in groups are smaller than

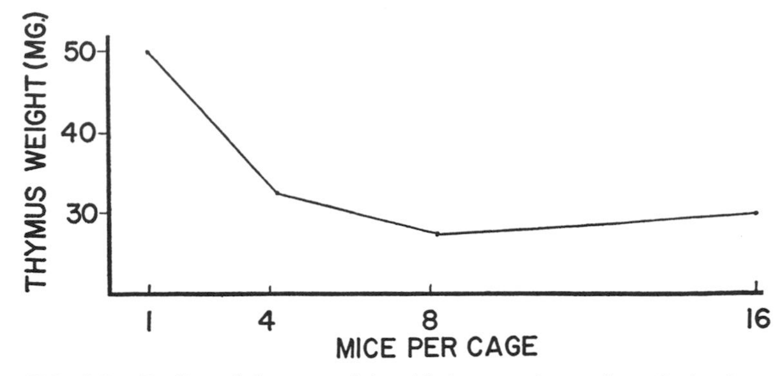


FIG. 3.8. Decline of thymus weight with increase in number of mice in a cage

those of male animals in isolation. For example, when mice in a growing population were compared with their controls in isolation, it was found that the testes in adults weighed 20 per cent less. Furthermore, an index of sperm production showed a measurable decrease in the grouped mice. Crowded female mice produced 5.4 viable embryos per pregnancy, but the controls had 6.2 viable embryos. The difference was not caused by a decrease in the number of ovulations but to loss of embryos after implantation. These effects on the gonads may result from an increase in corticoids or pituitary hormones or both.

Another effect of population crowding is interference with lactation. When baby mice born to females living in crowded conditions before conception are reared by their mothers, their growth is reduced in comparison to that of litter mates reared by foster mothers in non-crowded conditions. Whether quantity or quality of milk is affected is not known, but the reduction of growth lasts for at least two generations. Indeed, crowding inhibits growth in many respects. Young mice from crowded populations have testes weighing 84 mg., whereas the controls had testes weighing 143 mg. Other measures, such as growth of body and maturation of the ovary, also indicate retardation. Furthermore, activity may be reduced (Kelley, 1962).

Another action is the change in resistance to infection by pathogenic organisms. Corticoids inhibit inflammation (which protects against infection) and production of antibodies in many cases. Studies on mice show grouped animals to be less resistant to parasites (*Trichinella*) and to tetanus toxin. In the first case, inflammation is inhibited; and in the second, the level of antibodies is reduced. These effects also may be produced by corticoids and serve to increase mortality in high populations. An important point (though not a central theme of this chapter) is that aggressive behavior, acting through the adrenals, regulates the population by a feedback mechanism. As numbers increase, reproduction declines and mortality increases, thereby retarding the increase in population. (Thiessen and Rodgers, 1961).

All of these effects (and many others) can be experimentally obtained by injection of appropriate corticoids or ACTH. The evidence from the size of the adrenal, degree of involution of the thymus, inhibition of growth, and retardation of sexual development lead to the following conclusion: that aggressive behavior manifested in crowded populations stimulates the adrenal, through the pituitary, to secrete various corticoids. Recently, direct measurements of hormones from the adrenal vein support this conclusion (Nagra *et al.*, 1960).

The important point to be derived from the physiologic relationships is that corticoids are increased under conditions where aggression occurs. Aggressive behavior is thus recognized as one of many environmental variables (Arvay *et al.*, 1959) stimulating the hypothalamico-pituitary-adrenal axis. The sequence of responses are now documented for many situations. Any potentially harmful stimulus in the environment brings into action the general adaptation syndrome (Selye, 1957). For example, cold, poisons, noise, and behavioral situations will initiate the responses. Such a factor is called a "stressor" and is said to produce stress. The typical (and highly simplified) sequence of physiologic responses is as follows: first, the alarm reaction, then a stage of resistance or adaptation, and, finally—if the stressor persists—exhaustion and death. Of course, the stress may be so great as to produce death during the alarm reaction. The physiologic responses differ in the various phases. At first the available adrenal corticoids are exhausted and the catechol amines are secreted from the medulla; then the cortex slowly enlarges, producing more corticoids and the catechol amines return nearly to normal; lastly, after prolonged stress, exhaustion may occur. Many details of this scheme are controversial and require clarification.

Though many physiologic repercussions of aggressive behavior are known, knowledge of the pathways of action is meager. Stimulation by electrodes of certain areas in the brain results in an increase in corticoids (Setekleiv *et al.*, 1961), whereas in other areas stimulation has no effect. In particular, the amygdala is necessary for a response (Knigge, 1961). Removal of certain areas in the brain (Egdahl, 1961) results in an increase of ACTH and suggests that the brain tonically inhibits ACTH production. Furthermore, hormones from the posterior pituitary may directly stimulate the adrenal cortex (Hilton *et al.*, 1960; Smelik, 1960). An unsolved problem is the nature of stimulus to the hypothalamus. A probable stimulus is the secretion of hormones from the adrenal medulla. Innumerable studies (Elmadjian *et al.*, 1958) show that certain behavioral situations result in an increase of adrenalin. Some recent studies may be mentioned in detail. Levi (1961) reports that persons classified by psychologic tests as having high tolerance for stress had low levels of catechol amines. Frankenhaeusser *et al.* (1961) found that the stress of a psychologic task produced an increase in excretion of breakdown products of adrenalin. Further Melick (1960) examined the urine of medical students before, during, and after examinations. The excretion products of corticoids were significantly increased just before and during the examinations.

The catechol amines themselves may produce direct physiologic or pathologic consequences. A number of studies (see McNeil, 1959, pp. 260–64) show correlations between various states or stages of aggression and the occurrence of allergies (Bacon, 1956), constipation, headaches, ulcers, and many other symptoms. These consequences seem to be related to activity of the autonomic nervous system, which has well-known physiologic effects on blood pressure, glycogen in the blood, circulation, and other symptoms. Further, it should be noted that another result of catechol amines is the release of LH (Ginliani *et al.*, 1961), which (see above) stimulates aggressive behavior in some species.

Thus, we know that behavior, through a long chain of nervous and hormonal paths, produces an increase in two major types of hormones. We saw above that aggressive behavior produces, through the corticoids, many physiologic changes in reproduction and mortality rates. We also saw that stressful situations resulted in an increase of catechol amines. Some as yet inconclusive evidence suggests that these amines, perhaps acting through the posterior pituitary hormones, cause a release of ACTH, which, in turn, stimulates the adrenal cortex. While much work needs to be done, we have a possible chain of hormonal and nervous actions from the behavior to the physiologic consequences. But the behavior itself is associated with what humans call the "emotions" of fear and anger, which affect aggression.

## Emotions

The classic (James-Lange) theory of emotions states that an external stimulus in muscles and glands causes physiologic reactions which are felt as an emotion. Thus, one set of changes would be fear, and another set would be joy. Extensive research during the past fifty years has not substantiated the theory as originally stated. The theory is correct to the extent that emotions are clearly associated with hormonal and physiologic changes. Many studies show, however, that such changes can occur in the absence of emotions (see Morgan and Stellar, 1950). Experiments show that the same physiologic repercussions occur in different emotional states. The simple example of crying is pertinent. Persons cry for joy and for sorrow, yet the physiologic processes are the same. The difficult problem in the current state of knowledge is to determine which comes first, the emotion or the physiologic changes. Recent research shows clearly that the emotions may be separated from the physiologic processes by experimental manipulation. This fact has been apparent for a long time to persons who received injections of adrenalin. These persons had physiologic reactions (increased pulse, flushing, etc.) but did not feel angry. Similarly, extensive work on cats showed that manipulation of the nerves and hormones could produce physiologic responses almost identical to those of an angry cat when the cat was obviously not angry.

For our study of aggressive behavior we need to recognize that an animal responds to an attacking individual with physiologic responses that equip it to fight back or to flee. If attacks are persistent, the short-term reactions of the adrenal medulla are replaced by long-term changes in the adrenal cortex, resulting in its hypertrophy and in the long series of other reactions to corticoids. From the viewpoint of social behavior, however, one must consider how the animal first begins the responses. Schachter and Singer (1962) suggest that learning or cognitive factors are determinants of emotional states. Presumably a stimulus (which of course produced physiologic reactions directly) became associated with pain early in life. Later the sight, or even the thought, of that stimulus produced the physiologic reactions, which are now called "fear." Similarly, the same process, with many identical physiologic effects, established a connection between certain stimuli and joy. Thus, a pleasant situation may produce physiologic conditions identical to those produced by an unpleasant one. In the case of aggressive behavior, any emotions animals may have are presumably associated with their first experiences. In this way, the appearance of an attacking animal will become associated with pain or some other unpleasant feature and thus set the stage for fear.

Do animals other than man have emotions? This is one of the problems that have troubled students of animal behavior. Perhaps this problem will never be solved, but a reasonable approach is to explore the circumstances associated with the first attack. Perhaps subsequent exposures, through conditioning, will initiate the physiologic repercussions.

This chapter concludes by exposing a large area of the unknown in social behavior, namely, the kinds of stimulus and behavior which activate the physiologic repercussions to aggressive behavior and the manner in which learning assists its development. To encourage research and to summarize current knowledge, the following hypothesis is stated: Animals start life with an innate disposition to attack other members of the same species under certain circumstances. For this attack, the animal mobilizes, through the adrenal medulla and cortex, the bodily reserves and defenses. During the attack a number of stimuli (pain, sounds) become associated with the physiologic state, so that, these circumstances promptly begin to signify aggression. In this manner an attack becomes connected with certain physiologic responses, which, in turn, become recognized, by humans, as emotion. Eventually, merely the thought of the situation may stimulate the physiologic consequences.

## REFERENCES

Arvay, A., *et al.* 1959. Effect of severe nervous stimulation on the morphology and function of the adrenal cortex. *Acta. Physiol. Acad. Sci. Hung.*, **16**(4): 267–84.

Bacon, C. L. 1956. The role of aggression in asthmatic attack. *Psychoanal. Quart.*, **25**: 309–24.

Bevan, W., *et al.* 1957. Spontaneous aggressiveness in two strains of mice castrated and treated with one of three androgens. *Physiol. Zool.*, **30**(4): 341–49.

Birch, H. G., and Clark, G. 1946. Hormonal modification of social behavior, II: The effects of sex-hormone administration on the social dominance status of the female-castrate chimpanzee. *Psychosom. Med.*, **8**(5): 320–31.

Blauvelt, H. 1960. Personal communication (films).

CHRISTIAN, J. J. 1950. The adreno-pituitary system and population cycles in mammals. *J. Mammal.*, **31(3)**: 247–59.

———. 1963. Endocrine adaptive mechanisms and the physiologic regulation of population growth. *In:* R. G. VAN GELDER (ed.), *Physiological Mammalogy.* In press.

DAVIS, D. E. 1949. The role of intraspecific competition in game management. *Trans. No. Amer. Wildlife Conference,* **14**: 225–331.

———. 1957. Aggressive behavior in castrated starlings. *Science,* **126(3267)**: 253.

———. 1958. The role of density in aggressive behaviour of house mice. *Anim. Behav.* **6(3–4)**: 207–10.

ELMADJIAN, F., HOPE, J. M., and LAMSON, E. J. 1958. Excretion of epinephrine and norepinephrine under stress. *Recent Progr. Hormone Res.,* **14**: 513–54.

EGDAHL, R. H. 1961. Cerebral cortical inhibition of pituitary-adrenal secretion. *Endocrinology,* **68(4)**: 574–81.

FISHER, A. 1956. Maternal and sexual behavior induced by intracranial chemical stimulation. *Science,* **124**: 228.

FLICKINGER, G. L. 1961. Effect of grouping on adrenals and gonads of chickens. *Gen. Comp. Endocr.,* **1(4)**: 332–40.

FRANKENHAEUSSER, M., JARPE, G., and MATELL, G. 1961. Effects of intravenous infusions of adrenaline and noradrenaline on certain psychological and physiological functions. *Acta. Physiol. Scand.,* **51(2–3)**: 175–87.

GINLIANI, G., *et al.* 1961. Studies on luteinizing hormone release and inhibition. *Acta. Endoc.,* **38(1)**: 1–12.

GREELEY, F., and MEYER, R. K. 1953. Seasonal variation in testis-stimulating activity of male pheasant pituitary glands. *Auk,* **70(3)**: 350–58.

HILTON, J. G., *et al.* 1960. Vasopressin stimulation of the isolated adrenal glands: nature and mechanism of hydrocortisone secretion. *Endocrinology,* **67(3)**: 298–310.

KELLEY, K. 1962. Prenatal influence on behavior of offspring of crowded mice. *Science,* **135(3497)**: 44–45.

KING, J. A. 1957. Relationships between early social experience and adult aggressive behavior in inbred mice. *J. Genet. Psychol.,* **90**: 151–66.

KISLAK, J. W., and BEACH, F. A. 1955. Inhibition of aggressiveness by ovarian hormones. *Endocrinology,* **56**: 684–92.

KNIGGE, K. M. 1961. Adrenocortical response to stress in rats with lesions in hippocampus and amygdala. *Proc. Soc. Exp. Biol. Med.,* **108(1)**: 18–21.

LEVI, L. 1961. A new stress tolerance test with simultaneous study of physiological and psychological variables. *Acta. Endocr.,* **37(1)**: 38–44.

MACLEAN, P. D., PLOOG, D. V., and ROBINSON, B. W. 1960. Circulatory effects of limbic stimulation, with special reference to the male genital organ. *Physiol. Rev. Suppl.* **4**: 105–12.

MATHEWSON, S. F. 1961. Gonadotropic control of aggressive behavior in starlings. *Science,* **134**: 1522–23.

McNeil, E. B. 1959. Psychology and aggression. *J. Conflict Resol.* **3**: 195–293.

Melick, R. 1960. Changes in urinary steroid excretion during examinations. *Australasian Annals. Med.* **9(3)**: 200–203.

Morgan, C. T., and Stellar, E. 1950. *Physiological Psychology.* 2d ed.; New York: McGraw-Hill Book Co.

Nagra, C. L., Baum, G. T., and Meyer, R. K. 1960. Corticosterone levels in adrenal effluent blood of some gallinaceous birds. *Proc. Soc. Exp. Biol. Med.*, **105(1)**: 68–70.

Sawyer, C. H. 1959. Nervous control of ovulation. *In:* C. W. Lloyd (ed.), *Endocrinology of Reproduction*, pp. 1–20. New York: Academic Press.

Schachter, S., and Singer, J. E. 1962. Cognitive, social and physiological determinants of emotional state. *Psychol. Rev.* **69(5)**: 379–99.

Scott, J. P. 1958. *Aggression.* Chicago: University of Chicago Press. Pp. 149.

Scott, J. P. and Fredericson, E. 1951. The causes of fighting in mice and rats. *Physiol. Zool.*, **24(4)**: 273–308.

Selye, H. 1957. *The Stress of Life.* New York: McGraw-Hill Book Co. Pp. 324.

Setekleiv, J., Skaug, O. E., and Kaada, B. R. 1961. Increase of plasma 17-hydroxy-corticosteroids by cerebral cortical and amygdaloid stimulation in the cat. *J. Endocr.*, **22(2)**: 119–28.

Smelik, P. G. 1960. Mechanism of hypophysial response to psychic stress. *Acta. Endocr.*, **33(3)**: 437–43.

Snow, D. W. 1961. *A Study of Blackbirds.* London: George Allen and Unwin, Pp. 192.

Thiessen, D. D., and Rodgers, D. A. 1961. Population density and endocrine function. *Psychol. Bull.*, **58**: 441–51.

WILLIAM ETKIN

# 4

# Reproductive Behaviors

## Introduction

The sexual mode of reproduction is almost, though not quite, universal throughout the organic world. Even among the simplest of organisms such as bacteria and viruses which had previously been supposed to lack sexuality, the essentials of this mode of reproduction are now known to exist. Modern genetic theory expounds the evolutionary advantage of sexual reproduction over the asexual mode. It thereby explains the ubiquity of this type of reproduction, in spite of the obvious behavioral complexity which it necessarily involves in requiring the co-operation of two parents.

It has long been obvious that sexual reproduction allows for the mixture of hereditary materials from two different sources. It thus permits new characteristics of selective value which arise anywhere in a species to spread throughout the population. By bringing together favorable characteristics which have arisen by mutation in different parts of a population, sexual reproduction increases the variability within a species and furnishes a broader base for the operation of natural selection.

The role of sexual reproduction in promoting diffusion of genes is still regarded as important. But another and even more significant relation of sexuality to evolutionary efficiency has been brought to light by recent developments in the field of population genetics. In higher animals and plants, sexually produced individuals have two sets of genes. Such individuals are said to be diploid. If the members of a pair of genes differ (heterozygous condition), one of them is generally dominant, the other, recessive. The dominant gene is the only one which is expressed in the characteristics of the heterozygous organism. Since recessives are thus "covered" by the dominant gene in the heterozygote, natural selection cannot operate to eliminate them in such individuals. Only the homozygous recessive shows the recessive characteristics, and only in these individuals is the recessive gene subject to the action of natural selection. As a consequence, recessive genes, no matter how deleterious, accumulate in the gene pool of the population. The level of each recessive in the gene pool is determined by the balance between the mutation rate which is giving rise to this gene and the rate of elimination of the gene by the action of natural selection on the homozygotes produced when two heterozygotes mate. Each species population, therefore, contains a store of hidden recessives in its gene pool. Should the environment of the species change in such a way that previously non-adaptive recessive genes now become advantageous, the species does not have to wait for the slow process of mutation to originate the now advantageous genes, for they are immediately available in the gene pool. Because of the survival rate of the homozygotes under the new conditions, the proportion of the gene increases rapidly. Thus the evolutionary process is enormously accelerated by the sexual reproductive mechanism. We thus understand why, in spite of the numerous difficulties it entails, sexual reproduction is almost universal among animals.

## Patterns of Sexual Reproduction

### LOWER VERTEBRATES

The primary behavioral problem of sexual reproduction is that of bringing the sperm and eggs together to effect fertilization. The simplest method—one widely used among aquatic animals—is the scatter method. The gametes (sex cells) are simply released into the water, the random movements of the actively swimming sperm bringing them into contact with the eggs. Although in some plants chemical attraction is known to guide the sperm to the egg, no such attractive substances are known to act over a distance in animals. Many behavioral adaptations have been developed, however, which increase the probabilities of successful fertilization. The common tendency for animals to form aggregations during the breeding period or to grow in dense clusters, as do shellfish and tunicates on rocks, is one of these. A more direct adaptation is the almost universal limitation of spawning to one specific and limited time period. The swarming of the palolo worms two or three particular nights in the year, related to the lunar cycle, is a classic example of such limitation. Such behavior may be regulated by response to environmental cycles of tide, temperature and light, etc., or involve inherent time sense in the animals. In the worm *Bonellia* and to some extent in the marine snail *Crepidula*, the larvae which settle down near females tend to develop as males; otherwise they become females. The male of a deep-sea angler fish attaches itself to the genital region of the female, developing there as a degenerate parasite with little more structure than necessary for release of sperm. In starfish, worms, and some species of marine molluscs it appears that spawning is co-ordinated by the stimulating effects of the sex products themselves. Thus once a few animals start spawning, a chain reaction is set up whereby all the animals in the area are induced to release their sex products at the same time. Males and females or their sex products are brought into propinquity in numerous other ways, thereby facilitating fertilization even though the germ cells are randomly scattered.

Far more common and important, particularly in the lives of higher aquatic organisms such as fish, are the social behaviors by which insemination of the eggs is insured. As a result of behavioral interactions called "mating behaviors," males and females come into close proximity. They then release their sex products into the water in a co-ordinate manner so that the spermatic fluid (milt) is spilled directly over the eggs. Mating behaviors are complex. They involve co-ordination of seasonal development so that both sexes are ripe and active at the same, appropriate season. The sexually mature animal must be capable of recognizing the species, sex, and exact status of sexual readiness of a prospective mate. The mating procedure itself consists of a complex series of movements bringing the sex openings of the mates close together and effectuating release of the sperm over the eggs as the latter emerge.

The intricacies of this behavior can be illustrated in the analysis of mating in the common leopard frog (Noble and Aronson, 1942). First the males and then the females of this species emerge from hibernation early in spring with their gonads fully ripe, the development of the gametes having been completed during the previous summer. The males have a strongly developed clasping reflex and attempt to grasp any appropriately sized object that moves before them. When clasped by another, a male frog or a spent female croaks vigorously, a female distended with ripe eggs remains silent. The croaking serves as a stimulus for the clasping male to release; otherwise, as with a ripe female, the male remains in the clasping posture (amplexus) indefinitely. When releasing her eggs, which she normally does only while in amplexus, the female executes certain shuffling and pumping actions. These appear to stimulate the male to match her activities by releasing a dose of semen directly over each group of eggs as they emerge. When completely spent, the female shows backward shuffling movements. The male then releases his hold, and the mating is brought to an end. With an early spring start, the larvae have adequate time to complete their tadpole stage and metamorphose by mid-summer.

The extrusion of the sex products into the surrounding medium, common in water-living forms, is of course impossible in land-living animals, since the gametes would be quickly killed by drying in air. Thus internal fertilization is necessary in land forms. The sperm are introduced directly into the appropriate ducts in the female's body in a process called copulation or coition. Many animals have an intromittent organ (the penis in mammals) by means of which the sperm transfer is effected. Some accomplish copulation without specialized intromittent organs. Birds, except for ducks and a few others, lack penile structures and accomplish coition by a quick apposition of the lips of their external vents or cloacae. Bird copulation has been described as a "cloacal kiss." The male salamander in many species, after a brief courting of the female, marches ahead of her and deposits the spermatophore, a gelatinous capsule containing sperm. The female picks up the spermatophore with the lips of her cloaca. Still more bizarre are the reproductive behaviors of some of the cephalopods (octopus) which release one of their arms specialized to carry a load of spermatophores. This isolated so-called heterocotylized arm is seized by the female and inserted into her genital tract.

The advantages of internal fertilization in terms of conservation of sperm and eggs and certainty of fertilization, are manifest. Furthermore, it permits two other advantages: internal development of the egg and the secretion of heavy protective membranes around the egg before its release from the female. It is therefore not surprising to find internal fertilization widespread even among fish. The sharks and the many viviparous bony fish practice it. Transfer is usually accomplished with the aid of modified fins called claspers or gonopodia.

Internal fertilization as such does not necessarily entail any greater behavioral complexity than does external fertilization with mating as discussed previously. The protection and care of eggs and young, however, which occurs in both animals with external and those with internal fertilization, adds an entirely new dimension of behavioral complexity to the reproductive process. The care of eggs often involves nest building. In some fish this may consist of little more than clearing an area in which the eggs may be placed. The jewel fish, for example, merely cleans off a spot on a rock before placing her adherent eggs there. The nest of birds such as the gannets and penguins may consist of little more than a cleared area or "scrape." On the other hand, not only may birds weave elaborate nests, but fishes may also construct such extensive shelters. The male of many species of nest-building fish constructs the nest. The stickleback, a small fish breeding in the weedy shores of streams and estuaries, does this with sticks and weeds cemented together by mucous cords secreted by his kidneys. In birds it is more often the female that does the actual nest building (Fig. 4.1), but sometimes the male contributes or, rarely, as in the phalaropes, carries on alone. Completely domed nests, even provided with runways leading to separate outhouse conveniences are

FIG. 4.1. The crested cassique, a bird of the American tropics, lives in colonies in which each nest is a long pendulous sack elaborately woven of plant fibers and lined with leaves and other debris. The nest is built by the female while the male watches. She first wraps long fibers around a suitable branch and builds downward, working largely from the inside. When the nest is finished, copulation and egg-laying take place. Then the male abandons his family. (From an exhibit at the American Museum of National History.)

built by some birds (Pycraft, 1914; Burton, 1954; Heinroth, 1958).

From the point of view of behavioral correlation, nest building raises many problems regarding the nature of the activity. How is the timing of the nest building determined? It must, of course, come after territory establishment and before egg laying. How are the correlations between the sexes in their contributions to these activities effected?

With or without nests, animals may guard their eggs. The stickleback male keeps watch over his territory, including the nest with its eggs, driving off all intruders. But most of his time is taken up with fanning the eggs. This he does by directing a stream of fresh water over them with his fins. The rate of fanning increases with the age of the developing embryos and presumably keeps pace with their oxygen consumption. Parental care such as this is by no means rare among bony fish, patricularly those that breed near shore and maintain individual territories (Aronson, 1957).

Among amphibia and reptiles parental care is quite rare, though it does occur. The male of the midwife toad of Europe carries the strings of eggs wrapped around his legs. For the most part, he hides in damp places at this time but will emerge to dip the eggs occasionally in water when dryness threatens; he finally seems to recognize the appropriate time to bring them to a pond to permit the developed tadpoles to hatch. Though reptiles have made the very important step in the evolution of reproductive mechanisms by producing the first truly land egg, that is, one able to develop without external source of moisture, they are not in advance of lower groups behaviorally. A few reptiles show some parental care: pythons coil around their eggs and keep them slightly warmer than they otherwise would be, and alligators are said to protect and supervise their brood after they hatch. Parental care, however, only rises above the level reached by bony fish in the birds and mammals. These two groups are of sufficient importance to require individual treatment.

## BIRDS

Birds and mammals are warm-blooded (homoiothermal) creatures. Their bodies possess a thermostatic-regulatory mechanism controlling the production and loss of heat by the body. This mechanism maintains body temperature above that of the usual environment and keeps it constant despite fluctuations in the latter. The physiological advantage of homoiothermism lies in the greater metabolic activity possible at the higher temperature. The homoiothermal animal may continue its activity throughout the year, even in cold climates. But whatever its advantages, homoiothermism greatly complicates the problems of care of the young. The bird's egg, once it has passed an early stage of development, must be kept warm continuously by the parents; for, if chilled, the embryo perishes. This then requires incubation of the egg, a behavior which necessitates many adjustments. The parents must co-operate in some way to make it possible for at least one of them to incubate practically all the time. In pigeons, for instance, the female incubates from afternoon to the following morning, and the male takes over the rest of the time. Some passerine

**FIG. 4.2.** Post-nuptial continuation of courtship activities are common where both parents care for the young as in the gannets shown above. The gannets are seen with outstretched necks, clashing their beaks together like fencers crossing swords. This ceremony is repeated when one parent returns after being away on a feeding expedition. (After E. Armstrong, *The Way Birds Live*, 1943.)

and marine bird parents alternate in incubation for brief periods during the day, and one of them takes over at night. In some instances one parent (in the European robin, the female) does all the incubating. In any case, the problem of feeding the incubating parent remains. The male English robin brings food to his female, who leaves the nest long enough to partake of it. Where incubation is more evenly shared, food may still be supplied to the sitter by the mate, or each may forage for itself while off the nest. The parents must recognize and co-operate in relieving each other at appropriate times. Sometimes elaborate "nest-relief" ceremonies have developed in connection with the exchange of places at the nest (Fig. 4.2).

When the young birds first hatch, their feathering and physiological mechanisms for temperature control are not completely developed. Indeed, many birds (altricial) are hatched naked and quite unable to take care of themselves, and others (precocial), though feathered and able to feed and move about are delicate. It is essential for the physical protection of the young that incubation pass over into brooding. In brooding the parent tucks the young under her (or his) body and wings, thereby keeping them warm and also protecting them from rain or excess sun (Fig. 4.3). If naked young are exposed to the elements for too long (an hour may be too long in small birds such as wrens), they may die. Of course, effective brooding requires co-ordination between parents and young and between the two parents.

FIG. 4.3. The female in many birds as the red-backed shrike (*above*) protects the young from excess sun by shading them. The automatic character of this behavior in birds is illustrated by the fact that the female of one species has been seen to shade the nest even after the young were removed to the side. They then lay exposed to the sun in plain view of the female while she continued the shading behavior. (After E. Armstrong, *Bird Display and Behavior,* 1942.)

The brooding behavior is further complicated by the high food requirements and related factors. Young birds have an extraordinary rate of growth. The weight of young English robins was found to increase nine times in as many days. This high growth rate requires high food intake, in addition to the extra food necessary to maintain the homoiothermal condition, for energy. Many young song birds eat their own weight, or near it, in food per day. The feeding problem of a pair of song birds with a family of five is thus enormous. English robins have been seen to make, on the average, fourteen visits with food per hour, in one instance bringing over one thousand caterpillars to their young in one day (Lack, 1953). The young of other small birds (Fig. 4.4) were found to be fed three or four times per hour (Kendeigh, 1952).

Even after they are well grown and fully feathered, the fledglings are still a parental responsibility. Often the parents have to induce them to leave the nest, using "tricks," such as not feeding them, etc. But even out of the nest, the young are usually fed by the parents and only gradually become able to fend for themselves. During this period the parents, in so far as they are not preoccupied preparing another nest and brood, also guard the young, giving warning to them

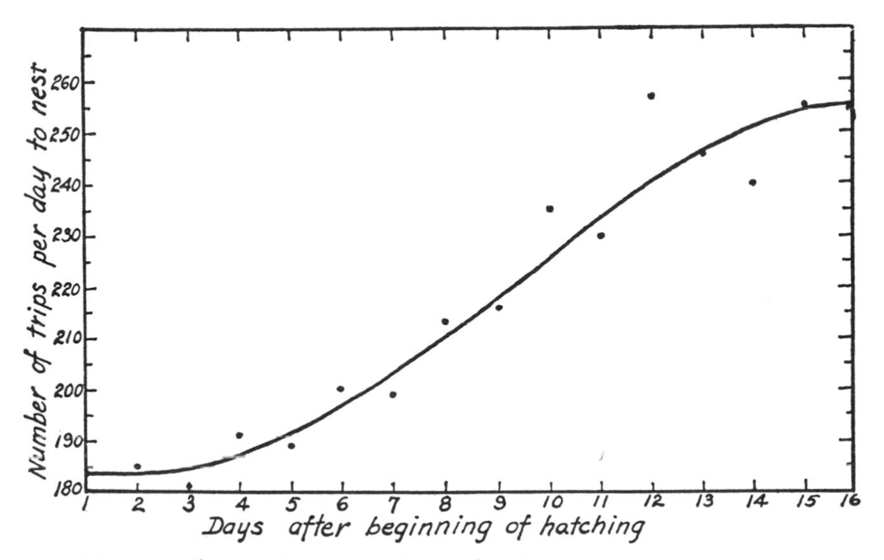


FIG. 4.4. The number of feedings the house wrens bring to their young increases as the young grow older and larger, requiring more and more food and less brooding. (After S. C. Kendeigh, *Ill. Biol. Monogr.* 22 [1952].)

by vocal signals of the approach of danger and in many cases attacking predators. Perhaps the most extraordinary parental protecting devices is the distraction display, or "broken-wing act," performed by many species of birds. When a predator approaches the nest or fledglings on the ground, one of the parents may rush in front of the enemy and flutter about on the ground with one wing awkwardly extended as though it were broken. As the predator is tempted to pursue this seemingly easy prey, the parent manages to keep just ahead of it and lead it away from the young, which meanwhile seek hiding places. Once while watching some newly fledged young robins hopping awkwardly on the ground, the author noted a cat approach-

FIG. 4.5. A common response of young birds to an alarm note sounded by a parent is to freeze in crouching attitude. Birds that show this response such as the stone curlews shown above are generally cryptically colored. As a result, such animals become almost invisible to the human, and presumably, to the predator's eyes. (Drawn from a photograph.)

ing stealthily. Rushing, as he thought, to the rescue of the young birds, he suddenly found an adult robin with broken wing fluttering frantically at his feet. Unthinkingly he reached to retrieve it, but it just eluded his grasp. Again and again for some twenty feet, he ran and reached for it, only to see the bird in the end dart off with perfect flight. Only then did he realize that he had been victimized by the broken-wing trick and led away from the young, which had meanwhile disappeared. We cannot now attempt to explain such behavior in objective terms but must be content to note it descriptively as an example of the extraordinary complexity of the behavioral adjustments between parents, both male and female, and their offspring (Emlen, 1955). (See Fig. 4.5.)

Compounding the intricacies of bird behavior is its co-ordination with seasonal conditions. In a general way, as we noted above for the frog, all animals breed only at appropriate seasons. But with the increase in complexity of reproductive life in birds, there is a corresponding increase in refinement of this co-ordination. Most birds of temperate climates migrate south when winter comes, and those which do not must be prepared for a period of great hardship in winter. An early start and a rapid growth rate during the first summer of life are of great advantage in preparing birds for their first winter. Yet, if the parents breed too early, there may not be enough insects or other food for the voracious young when they hatch. Consequently many birds must follow a precisely regulated schedule. Their complex pairing, nest building, copulation, egg laying, incubation, brooding, and feeding of young must not only be attuned to the regular cycles of the seasons but also subject to some further control to fit the vagaries of the weather in each season (Fig. 4.6). The northward-migration dates of many birds are quite regular to the calendar, some arriving on almost the same day year after year, and most of them showing but small variation in time. For example, in song sparrows near Columbus, Ohio: the first breeding males were found to arrive within a week of March 1, but the day varied within this period in conformity to the average mean temperature of the last ten days in February. Similarly, the first eggs were layed between April 15 and 23, and, within that period, laying was closely related to previous average

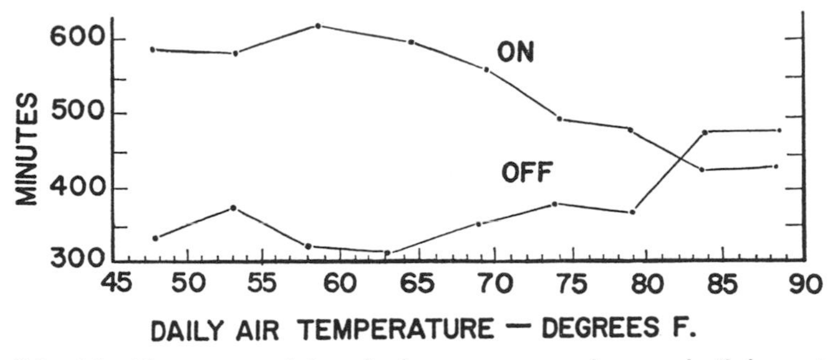


FIG. 4.6. The amount of time the house wren spends on and off the nest varies with the temperature, as shown above. Thus, incubation is much more persistent at low temperatures than at high. Above 80°F. the female spends more time off the nest than on it. (After S. C. Kendeigh, *Ill. Biol. Monogr.* 22 [1952].)

temperature (Nice, 1937). Even after being started, nest-building and other reproductive activities were interrupted or slowed down by a spell of bad weather. It is clear that in birds there is an overall mechanism that somehow regulates these behavioral activities in accordance with the season of the year; minor mechanisms further fit the behavior to the weather conditions of each particular year. All in all, the behavioral adjustments between female, male, young, and the environment in birds are so complex that they probably constitute the most highly elaborated reproductive behaviors found anywhere in the animal kingdom. Of particular interest for our later discussion is the fact that the successful raising of young birds depends upon long-continued co-operation between male and female. In song birds the sequence of territory establishment, pairing, nest building, mating, egg laying, incubation, brooding, and feeding of young is accomplished sometimes by one parent, sometimes by the other but in most species is shared by both.

### MAMMALS

The basic biological organization of mammals is, in some respects, simpler and, in others, perhaps more complex than that of birds. We shall be concerned here only with the highest group of mammals, the placentals, since this includes almost all of the familiar mammals. The other groups, the egg-laying mammals and the marsupials, are but a surviving remnant of early types now largely confined to the Australian region. Placental mammals are characterized by the fact that the embryo develops within the mother's uterus, deriving nourishment and oxygen from the mother and giving up its wastes to her through a temporary organ called the placenta. At parturition (birth of young) the placenta separates from the uterus and is eliminated as the after-birth, which is generally eaten by the mother.

Though all placental mammals are born in post-embryonic stages, that is, with completely formed organs, there is considerable variability in the degree of development. Some, such as the bears and the primates including man, produce relatively small and helpless young that require many months of nursing and care before they can move about and find food for themselves. Others, such as the hoofed animals, porcupines, guinea pigs, and hares, are relatively far advanced in development at birth. Buffalo young can move along with their mothers within a few hours after birth, and hares can feed on solid food within a few days. Young porcupines become independent of their mothers in a week or so (Bourliere, 1954). In contrast to birds, many mammalian young develop quite slowly. In the larger land mammals nursing periods of a half to a full year are common. Bear young do not become independent of their mothers for two years. Mice require about three times as long as birds of comparable size to attain adult weight. In most ungulates sexual maturity is not reached for two or three years; in chimpanzees, for six to eight years; in humans, for about fifteen years; and in the elephant, not until about thirty years. During the nursing period the young are dependent upon only the female parent for food and general protection. Thus the initial dependence of the young centers around the female parent and, in contrast to birds, generally contains no role for the male. The long prepuberal developmental period affords extensive opportunity for learning and for the development of social relations between the mother and offspring.

The integration of the mammalian reproductive pattern into the seasonal and other environmental variables likewise is centered chiefly around the female. The larger mammals, particularly in temperate or arctic climates, have only one brood per year or sometimes one in two years (bears, for instance). The females and males are sexually active only during a limited period, the "rut" period. The males become active early in rut and maintain a more or less continuous sexual activity throughout it. In some mammals sexual activity is maintained throughout the year, and in many domesticated forms the limited rut period natural to the species tends to become continuous throughout the year. In such cases the males remain sexually active continuously during adult life.

In females the situation is more complex. There is cyclic activity, called the estrus[1] cycle, which consists of short-termed physiological cycles containing a brief interval of sexual excitement called heat or estrus. These cycles go on during the rut period but cease during the rest of the year. Thus sheep and most large ungulates show a fall rut period and are sexually quiescent at other times. In contrast to the males, which are continuously active during rut, the females accept the males only during their heat phase of their estrus cycle; if they are not impregnated at the first estrus, the cycle may be repeated at regular intervals during rut. In the non-rut period, estrus cycles cease. Such mammals are said to be seasonally polyestrus. Many tropical and domesticated mammals and some wild, temperate climate species run cycles all year long (permanently polyestrus). A few, such as bears and seals, have only one estrus period during the rut season (monestrus).

Some aspects of the estrus cycle must be mentioned briefly because they are important for an understanding of mammalian behavior. The term estrus (older spelling, oestrus) previously meant the period of sexual receptivity or heat in the female. Today, however, this is known to be correlated with changes in the entire reproductive system of the animal, and the term has been extended to include these changes as well as heat behavior. In a typical mammal, such as the guinea pig, the follicles containing the eggs ripen in the ovary during the early period of the cycle (proestrus), and the uterus undergoes certain

[1] The spelling "estrus" is used here for both adjective and noun.

changes. In estrus proper (the period of heat) or a few hours thereafter, ovulation takes place and, if coition has also taken place, the egg is fertilized in the upper part of the female ducts. In the next phase of the estrus cycle (metestrus), the embryo implants itself on the wall of the now fully prepared uterus. A placenta is established and pregnancy results. Should implantation of the egg fail to occur for any reason, the animal passes into a sexually quiescent stage (diestrus). After the diestrus rest period the animal runs another estrus cycle during the breeding season. If the non-breeding season meanwhile supervenes, estrus cycles cease altogether (anestrus). The rabbit and the cat and certain other mammals do not automatically run through the estrus cycle but remain in heat for some time until mated. During pregnancy the cycles generally cease except for traces. Many mammals also run one cycle after parturition, but during the period of lactation, cycles again cease. In some species even after an infertile copulation the female remains in a pregnancy-like state called pseudopregnancy, during which cycles are suppressed.

When the primate uterus which had been built up in preparation for an embryo breaks down after metestrus because of the failure of implantation, there is considerable loss of blood as the uterine lining is cast off. This is called menstruation. As can be readily understood from the origin of the bleeding, menstruation takes place not at estrus but rather between succesive heat periods in about the middle of the estrus cycle. In most primates the time of estrus can be identified by behavioral receptivity of the female and by structural changes in the sex system. It can thus be seen to alternate with menstruation. In the human, however, though menstruation is conspicuous, there are no definite estrus behaviors or morphological changes by which the time of ovulation can be observed, although it can be detected by refined electrical and temperature recordings. In summary, we may say that most mammals show estrus cycles characterized by definite heat behavior; primates, other than man, show both estrus and menstrual cycles; and man shows only menstrual cycles. Most of the fundamental physiological changes are similar in all three types. In a few lower mammals some bleeding from the genital tract may occur which is not menstruation. Dogs bleed slightly during estrus proper.

The nature of mammalian reproductive physiology as outlined above makes it clear that the burden of reproductive behavior necessarily falls primarily upon the female. It is her estrus behavior which determines the time of copulation, and she does not ordinarily permit

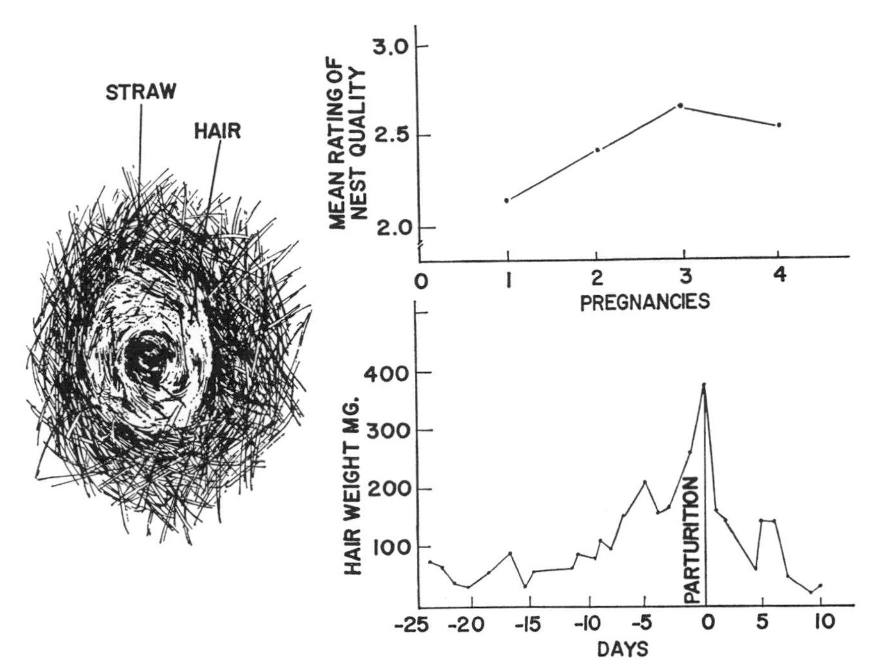


FIG. 4.7. Nest-building in rats is strongly affected by environmental temperature and by pregnancy. Both low temperature and closeness to the time of parturition increase nest-building. However, previous experience does not seem to be an important factor in these animals. Rabbits, on the contrary, improve the quality of their nest-building with the first three successive pregnancies, as shown in the graph, *upper right*. They line their nests with hair (*left*). This is pulled from the abdominal region where it becomes loose in the pregnant female. It is possible to quantify the changes associated with pregnancy by weighing the amount of hair which comes loose during a standardized combing procedure (graph, *lower right*). In the *left* picture the hair lining in the nest completely covers the young. (After Zarrow, Sawin, Ross and Denenberg, *in* E. L. Bliss, *Roots of Behavior*, 1962.)

copulation except in estrus. It is she alone who carries the burden of the developing embryo, and it is she with whom the young necessarily associate closely during lactation. If a nest or lair is prepared, the female builds it (Fig. 4.7). Only in a few species, as in the fox and the wolf, does the male contribute toward the care and feeding of the young. Generally, it is with the female that the young are associated, and from her they derive the basic mammalian social behaviors.

In contrast to the situation among birds, the contribution of the male mammal tends to be limited to that of insemination, and his social role is usually much reduced. His co-operation is generally not essential for the care of the young subsequent to fertilization. He commonly does not associate with the female and her young in any permanent manner. In some mammals, such as the cat and bear families generally, he is excluded by the female from contact with the young, being treated as their most dangerous enemy. We shall see that many of the behavioral characteristics of the mammalian male are related to the basic biological position of dispensibility after insemination. Of course there is much variability among mammals, but it is only exceptionally that the male finds an important niche in the social relations of parent and child among mammals. From the point of view of natural selection, mammalian males may be characterized as expendable. This characteristic helps to account for the kinds of specialization in behavior and structure that we find in many mammals (Bourliere, 1954).

## Mating and Courtship Behaviors

### GENERAL CONSIDERATIONS

The insemination of the female by the male, especially in land forms, often involves complex behaviors upon the part of both mates. Behaviors that directly lead to the transfer of sperm from male to female are called "mating behaviors." The commonly recognized behaviors of mounting, pelvic thrusts, and intromission of the penis in mammals or the "cloacal kiss" in birds are clearly to be included in mating behaviors. But in addition to these, we very commonly see many other activities between mates that do not seem to be directly involved in mating. For example, a male tern brings a fish to his "intended" mate. At this time the mate is perfectly capable of fishing for herself and this attention seems superfluous. After toying with the fish for a while the female may discard it. Clearly, hunger played no role here. Yet all along the shore in the spring, terns may be seen presenting pieces of fish to others and doing it, furthermore, with elaborate, precise, and formal ritualistic movements. Such activities between mates which do not directly help in insemination we shall call "courtship behavior." Of course, the distinction between mating and courtship behavior is arbitrary, and many behaviors are difficult to classify as either, since mating and courtship grade into each other. For example, during mating the male cat seizes the female by the scruff of the neck with his teeth. Is this behavior directly or indirectly involved in insemination? It would be pointless to try to distinguish mating from courtship here.

We will not, in fact, find it useful to try to maintain a distinction but will discuss both together in this section.

It is to be noted that the word "courtship" is used here in a much broader sense than is common in speaking of human behavior. For one thing, it is not confined to the male sex but may be conspicuous in female behavior. In the second place, animal courtship may continue after mating and be part of the characteristic activities of the two parents toward each other. Terns, for example, when they come back to the nest to relieve their mates in incubation, commonly bring a fish and present it with much bowing and posturing. Since this is clearly much the same behavior as took place when the animals were pairing up, we can call it post-nuptial courtship.

By the very terms of our definition, mating behaviors serve a clear-cut, useful function in reproduction. They enable the male to inseminate the female. We can, therefore, readily understand them in terms of adaptive behavior. On the other hand, courting behaviors are often not only bizarre, wasteful, and useless, but, as with the snarling and "fighting" of cats, seem to interfere with the smooth progress of the mating process. Yet they are common and conspicuous parts of the reproductive behavior of many vertebrates. Why?

We will not attempt here to describe even a sampling of the great variety of courtship behavior to be found in nature. Though fascinating, this would be a confusing and endless task. Instead we will attempt to analyze the principal functions accomplished by courtship behaviors and illustrate these by appropriate examples.

### FUNCTIONS OF COURTSHIP

*Advertisement.*—Clearly one of the problems in mating is the recognition of appropriate mates not only as to sex but also as to readiness to mate. Behaviors that "advertise" the sex and readiness of the mate constitute one class of courtship. Whereas the male of many vertebrate species is continuously sexually active during the reproductive season, the female is suitable as a mate only during a limited estrus. Therefore, the estrus female must provide the clues for mating to the male. Many of these are physiological rather than behavioral in nature. Secretions produced by the female in one fish were found to stimulate mating behavior in the male (Tavolga, 1956). Since most mammals are primarily nocturnal, non-visual animals, they use the sense of smell widely. Many female mammals in heat can apparently be recognized by odors, sometimes from special anal glands like those of the cat. Male rats distinguish estrus from non-estrus females, even at some distance, by odor, and male dogs are attracted to the urine of the estrus bitch. In female chimpanzees and some other female primates the estrus state includes swelling and reddening of the so-called sexual skin in the perineal region. In addition, the estrus female in mammals generally shows behavior that differs from the non-estrus condition. The non-estrus female is antagonistic to the approach of the male, avoiding or repelling his advances. As the minimum of estrus behavior then, the female modifies this repulsion so that she stands still for the male and tolerates his sniffing and eventually his mounting behavior. In many of the larger ungulates, such as deer, estrus behavior shows little more than this standing for the male. The female shrew will kill the male confined with her in a cage except when she is in estrus. Many rodents, rats, guinea pigs, etc., when stimulated by a male during estrus, show a special posture response called "lordosis." In this, the back is arched downward in the center, bringing the genital region up, and the tail, if present, is turned sharply to one side. This lordotic posture, of course, directly assists in permitting the male to achieve intromission. In fact, intromission cannot be achieved without such co-operation in these or in other mammals with the exception of man and perhaps some higher primates. In addition to lordosis, the female rat in heat shows a peculiar start-and-stop running when followed by a male. She is also characterized by a tremendous increase in running activity, which presumably facilitates contact with the male under natural conditions. The female lion not only exposes herself to the male but will directly crawl up to and squeeze underneath him if he is not otherwise responsive to her advances and odors. These advertisement displays of the female in estrus are, however, not a particularly elaborate aspect of courtship in vertebrates, because, as mentioned before, the male generally maintains sexual receptivity continuously during rut and actively seeks the female. The biological "need" for female "advertising" display is not great. Nevertheless, the female vertebrate generally does display a definite estrus period of receptivity even if, as in the guppy, it is so subtle that it is difficult for the human observer to identify it (Clark and Aronson, 1951).

Advertising displays of the male are a more conspicuous, though perhaps not more common, aspect of animal courtship. We appreciate this best in birds, lizards, and some fish which, like the human, are "visual-minded." The conspicuous "showing off" of fine feathers by the cock of the barnyard fowl and of many other birds is well known. Because in its extreme form this type of display seems more related to the correlation function of courtship than to simple advertisement, it will be discussed under this topic later. Here we may point to examples that more simply fit the concept of advertisement function. The male of the common heron of Europe builds a partial nest in the tree tops of the colony area. This is readily seen from above. He then

FIG. 4.8. In the Jackson Whydah, a bird of the plains of Kenya, many males establish their "courts" together. Each consists of a ring on the ground, about ten feet in radius, trodden flat around a central tuft. In this "court" the males advertise for females by fluttering up and down like "Yo-Yo toys." When a female lands in the arena the male raises and quivers his neck hackles and tail plumes and occasionally flounces his tail in the female's face. (After E. Armstrong, *Bird Display and Behavior*, 1947.)

places himself in this nest foundation and assumes a very awkward and conspicuous posture, erect with neck stretched, head pointing upward, and neck feathers fluffed out. At the same time, he calls with a loud sharp cry. This strained behavior may continue for many hours a day for many days until a female flying by is attracted and stays to mate (Stonor, 1940). For lek birds, such as the ruff and sage grouse (see Chap. 1), in which many males take up small mating territories close together, the essential function of the group display of fine feathers and noise would appear to be the advertisement of a group of males ready to mate (Fig. 4.8). Females in the estrus state can then readily find suitable mates (Armstrong, 1947). The male of many territorial mammals, in marking off his territory (by odors as the dog does with his urination stations or by sound as the moose does by bugling), not only notifies other males of his territorial rights but presumably communicates relevant information to the females. The loud singing of territorial passerine birds likewise serves to alert females to an available territory. The posing and flashing of his brightly colored gular-skin flap by the male anolis lizard in his territory has been found to attract females from afar (Greenberg and Noble, 1944). Similarly, the patrolling of territory boundaries by conspicuously marked males in fish such as the sticklebacks also serves an advertising function.

*Overcoming of aggression.*—On the whole, the advertising function of courtship accounts for only the simpler aspects of courtship. Another function which courtship serves is that of overcoming the aggressive responses of one animal to another. Such aggression may be considered here under these three headings: the aggression of predators, of dominant animals, and of territory owner.

(*a*) Predators: The behavioral mechanisms of a predaceous animal, of course, prompt it to attack any other animal approaching it. It is clear that if male and female are to mate successfully, the predaceous behaviors of the mates must be overcome by some signalling system. This task is often accomplished by conspicuous courtship behaviors. Perhaps the classic example of this is seen in various species of jumping spiders, which feed by pouncing upon their prey. Since the female is larger than the male, any male approaching a female is in danger of ending up as food rather than as mate. In such species the males approach the females with elaborate courtship dances. While still a safe distance away, the male assumes unusual postures rising high on its legs and waving conspicuously marked appendages before the female. Such dancing may last for hours, and mating is not consummated unless the female shows appropriate signs of quiescence (Fig. 4.9). Some predaceous flies wrap a victim in silk and present this to the female as part of the courtship; and, according to the usual interpretation, while she is occupied with this "present," mating is successfully accomplished. Still more bizarre actions have been observed, such as the female spider going into catalepsy before the male makes his final approach or submitting to being tied firmly to the ground with silk spun by the male (Bristowe, 1941).

Solitary predaceous mammals, particularly of the cat and weasel family, often do considerable fighting as part of their courtship. The

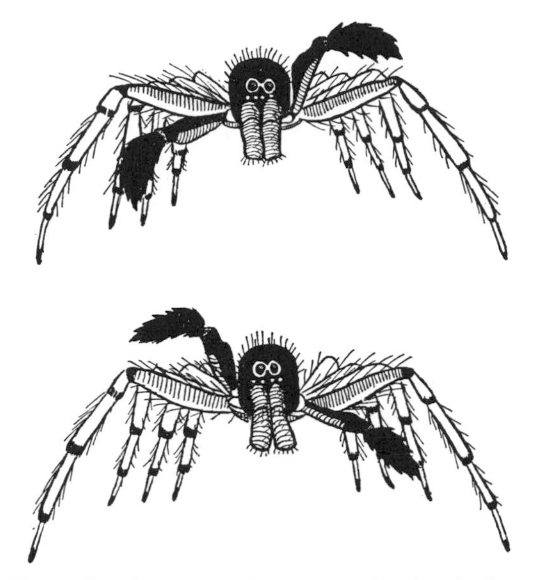

FIG. 4.9. The male of some spiders courts the female by waving his conspicuously banded appendages. The female is induced to go into a trance-like state during which the male safely mates with her. (Redrawn after Bristowe, 1941, and M. Burton, *Animal Courtship,* 1953.)

mating of cats is accompanied by much scratching, biting, and appropriate vocalization. This behavior differs from true fighting, for only minor injuries arc inflicted; and since neither animal yields, it continues for some time. The mating process of ferrets and mink looks like a prolonged fight, often lasting an hour or more. The courtship fight arouses the animals to a high level of emotional tension which, as we shall see later, has important physiological consequences. Behaviorally, it appears to be fighting behavior normal to a predator but modified and controlled by stimuli emanating from the sexual situation.

(*b*) Dominants: The aggressive behavior of group-living animals is, as we have seen in Chapter 1, subjected to patterned control by the formation of dominance hierarchies. The development of mating behavior between males and females often involves important modification in these dominance-subordination patterns. In groups such as those of the baboon or the Indian antelope where the male overlord is supremely dominant, the other members of the group pattern their movements in relation to his, keeping out of his way, permitting him to take over any desirable object without dispute and yet remaining near —but not too near— him. When a female baboon comes into heat, however, she comes close to the male, standing her ground when he makes aggressive passes at her in a way which she never does when not in heat. She also displays her genitals in the so-called presentation posture. In this pose the animals faces away from the dominant, bends over at the waist in such a way as to expose the perineal region to the view of the male, meanwhile watching him with sharply turned head. At this time her sexual skin is engorged and highly colored, revealing her estrus state. This posture affords maximal convenience for the male to mount her (Zuckerman, 1932). Presentation is a common sign of submission to a dominant in many primates and is adopted by male subordinates as well as female. Thus courtship behavior in the female monkey consists principally of subordination, sexual receptivity, and acceptance of punishment from the dominant male. Male dominance is mitigated to some extent, since the punishment administered to the estrus female when she violates his dominance privacy is relatively mild and soon stops altogether (Carpenter, 1942). It has been shown experimentally that the female chimpanzee becomes more assertive toward her mate, often assuming a typically dominant acquisitiveness during estrus. (Birch and Clark, 1946) Alteration in dominance at estrus has also been recorded for many animals. In some birds, such as the budgerigar or parakeet, the female generally dominates the male except when she enters estrus. A female pigeon has been observed to rise in dominance position with respect to other females upon pairing with a dominant male, and similar interactions of dominance and mating associations has been reported in the Japanese monkey (Imanishi, 1957).

(*c*) Territory Owners: One of the commonest forms of aggressive behavior, as we have seen in Chapter 1, is that connected with territoriality. The common form of individual territory in vertebrates is that set up and defended by the male. Obviously, if mating is to be achieved in such forms, the female must somehow gain access to the territory and achieve a *modus vivendi* with the male. The modifications in behavior which are involved and which we here consider to be courtship may be centered around three main techniques: (1) display of morphological sex recognition marks, (2) behavioral characteristics marking the female as distinct from the male, and (3) display of subordination behavior.

When a stickleback swims into a male's territory, the owner swims to attack. If the other animal is a male or an unripe female, it generally swims away. If the invading male is "inclined" to dispute the territory, he displays his red belly by turning vertically upward with his ventral side toward his opponent. A ripe female, on the other hand, turns up but reveals the egg-swollen silver belly instead. The territoral male's reaction to the male display is aggressive, but the display signals from the ripe female modify this aggression so that the territorial male turns and swims toward the nest. He turns again to attack, reverts again toward the nest, etc. This zigzag dance that thus characterizes the courtship is a modification of the aggressive behavior shown toward a male (Tinbergen, 1953). (See Fig. 8.4.)

The courtship behavior of the American song sparrow likewise is built around territorial aggression modified by the nature of the female reaction. If the male displays, sings, or even pounces on the female, instead of the female replying in kind as a male would, she persists in the territory without fleeing or showing aggression. Because of this receptive behavior, the male's display of aggression is soon diminished, and he becomes reconciled to her sharing his territory (Nice, 1937). This type of behavior is seen in many territorial birds. Even the heron, which goes to such lengths to display his availability as a mate, attacks the female with fierce determination when she finally does come to his nest. A primary factor in achieving the pairing is the ability of the female to accept the initial aggressiveness without fighting back. Since many birds are monomorphic, that is, male and female are alike, this subordination behavior seems to be the chief sign by which the sex of the individual is recognized. On the other hand, if the sexes differ morphologically, this serves as a means of sex identification. The male flicker has a mustache-like arrangement of black feathers lacking in the female. After some black feathers were attached to a trapped female to form a "mustache," she was attacked by her mate when she was permitted to fly back to her nest (Noble, 1945).

*Physiological co-ordination of male and female.*—Successful reproduction requires as a minimum the co-ordination of male and female activity with respect to insemination. In many vertebrates, most particularly in birds, the co-ordination must extend far beyond that of insemination. In some, both parents co-operate in a long process beginning with territory establishment, continuing through nest building, copulation, egg laying, incubation, brooding, and feeding of the young until they are ready to leave the nest. It is not surprising, therefore, to find in such birds courtship activities that are not only elaborate but extend through the entire reproductive season. Shore birds such as gulls, terns, and gannets provide excellent examples. Herring gulls, for example, pair before choosing territories (Tinbergen, 1953). Many birds gather in a separate area called the "club." Here a female may approach a male and circle him with subdued mien and calls. His responses include aggressive threats chiefly directed to neighboring males, scraping and choking movements such as characterize nest building, and finally regurgitation of food. The regurgitated food is eagerly seized by the female. Such activities may go on for days before the pairing bond is definitely established. In a well-established pair that has taken up territory in the breeding colony, these same movements—especially choking, food begging, and regurgitation feeding—are continued. Now, however, they are a mutual activity that continues between mates, though the female does not actually feed the male when he begs. Here we see courtship that includes submission signals and activities that are reminiscent of later stages of reproduction, such as nest building and feeding of the young. Though modified in form and intensity as the season progresses, some of these activities continue as mutual social exchanges between mates. Many reproductive activities are mutual. Both mates show nest-selecting activities by making scrapes here and there; both collect nesting material; both incubate and feed the young. The behavioral contacts provided by courtship activities appear to play the role of co-ordinating the partners by stimulating the slower one to reach the higher level of its partner. When a bird returns from a feeding trip, the mate will usually rise from the eggs and permit it to take over incubation. If, however, the mate does not respond, the returning bird will call, show choking behavior before it, and fetch nesting material to present to it. These activities, which are called "nest-relief behaviors," are common in birds and may be considered part of the post-nuptial courtship that helps to co-ordinate the activities of the pair. Sometimes nest-relief ceremonies are as striking and formalized as other courtship behaviors. Gannets, for example, show the same breast-to-breast posturing with outstretched neck and clashing bills at nest relief as in earlier courtship behavior. Such behaviors, posturings, nest-material exchange and real or "token" feeding, are commonly maintained by pairs of birds throughout the reproductive season. They can only be regarded as behaviors helping to maintain and synchronize the pair in their mutual activity (Huxley, 1914). Examples of such mutual courtships are shown and discussed in Figures 4.2, 4.10, 4.11, 4.12, and 4.13.

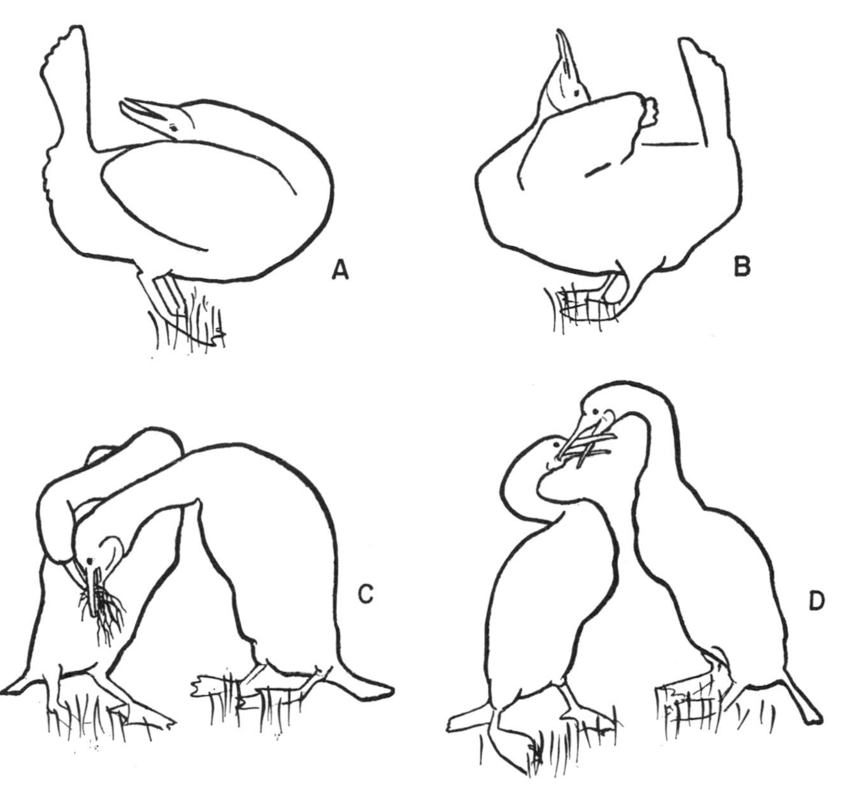


FIG. 4.10. The courtship of Brandt's cormorant, like that of so many other aquatic birds, involves much mutual stimulation. The female solicits the male by assuming an extraordinary posture with neck twisted over her back and tail erected (*A*). The male assumes a somewhat similar posture in courting the female, thereby displaying the vivid blue pouch under its neck (*B*). After mounting, the male brings nesting material to the female and together they place it in the nest site (*C*). They do not feed each other as so many other species do but nibble and grip each other's bills, twisting and swaying together (*D*). The comparability of male and female behavior extends as in other birds to the occasional reverse mounting of the male by the female. (After E. Armstrong, *Bird Display and Behavior*, 1947.)

Since both male and female song birds are confined to the same territory during reproduction, the animals necessarily maintain behavioral contact and thus facilitate co-ordination. Courtship persistence during later reproductive phases, though occurring to some extent, is certainly not as conspicuous here as with marine birds, where it is functionally more important.

Shore birds and song birds are generally monomorphic; that is, the sexes are similar. Even an expert finds it difficult or impossible to dis-

FIG. 4.11. Courtship in the great crested grebe involves elaborate mutual ceremonials. Some incidents of these activities are illustrated above. At *A* we see the head shaking ceremony in which the pair face each other displaying their head feathers and shaking their heads from side to side. *B* shows the male dive in which the male approaches the female with head submerged and suddenly shoots high out of the water just in front of her. At C we see the mutual presentation of water weeds to each other. (After J. S. Huxley, *Proc. Zool. Soc.,* 1914.)

tinguish male from female without behavioral signs in gulls, song sparrows, etc. Some of the most striking, and certainly the most conspicuous, types of courtship occur in those animals in which male and female are dissimilar, i.e., dimorphic species. In these species, it is almost always the male that is the brightly plumaged sex. The barnyard rooster, the turkey cock, the bird of paradise, and the peacock are well-known examples. Pictures of their fancy feathers, however, give only a faint idea of the true nature of their courtship. Not only do they spread these showy feathers before the female, but they do so in ways that arouse a maximum of interest. The turkey cock, for example, parades a side view with one wing dropped and the spread of tail feathers twisted toward the female. The peacock not only spreads his gorgeous fan of back feathers but rustles them as well, sometimes opening the fan while facing away from the female and then suddenly wheeling around with a great rustling. The effect upon a human observer, at least, is dazzling and exciting. In many displays, as of some of the birds of paradise and their relatives, the bird assumes grotesque attitudes in which all resemblance to a bird is lost.

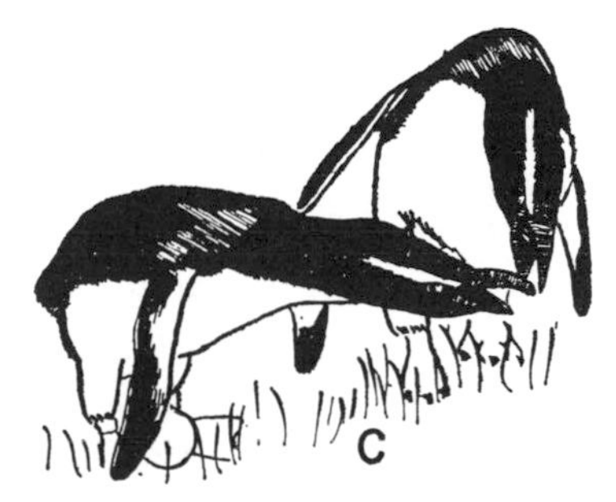


FIG. 4.12. The yellow-eyed penguin often courts the female with the posture called the "salute" (A). This same posture is often shown as a social greeting between any two animals and is not restricted to courtship activities although it appears to be common then. The "ecstatic" posture of the erect crested penguin is a mutual ceremony in which a pair indulge during pair formation and afterward during post-nuptial courtship, when one individual relieves the other in incubating the egg. First the birds pose upright (B), then bow deeply, rocking from side to side and calling loudly (C). (After L. E. Richdale, *Sexual Behavior in Penguins,* 1951.)

FIG. 4.13. The courtship dance activities of the wandering albatross begin with mutual nibbling of the beaks (*above*) and reaches a climax with the ecstatic display of the huge wings by both male and female. (Drawn from a photograph.)

If sounds accompany the display, they are loud and raucous rather than melodious. All in all, such displays clearly tend to arouse a high pitch of excitement in the female. The functional significance of these displays must lie in the stimulation of the female to the mating act. As we shall see in the discussion of the physiological correlates of these phenomena (Chap. 6), there is much evidence that such courting excitements stimulate the reproductive mechanisms of the female.

The effect of a single displaying animal is exaggerated when the display is part of a communal activity (Fig. 4.14). Even in non-colonial birds, sexual excitement is contagious (Collias and Jahn, 1959). In fish the turmoil of breeding aggregations seems to have stimulatory value (Aronson, 1957). The excitement at a ruff lek when a female appears reaches a high pitch with each male bounding up and down in his tiny territory like a Ping-pong ball and assuming ecstatic attitudes with the colored ruff feathers extended fully when the female is close by. Such communal excitements are believed to be achieved even in the much less garish displays of the communally nesting shore birds. In the gulls, there is evidence that the communal nature of the display is important in correlating the phases of activity in all members of the colony (Darling, 1952).

Since the estrus cycle is the fundamental co-ordinating mechanism in mammals, it would seem that behavioral co-ordination by way of courtship would not be as essential. In general this is the case, and mammalian courtship is much less elaborate than that of some birds. Yet the courtship of mammals also produces great excitement. Many carnivores show a rough and tumble courtship in which fighting is prominent. In the cat and weasel families, the female does not ovulate spontaneously during estrus but remains in heat until mated. Then the excitements of the mating behavior lead to ovulation. In ecological terms, this would appear to be neatly adaptive, since in such solitary carnivores the female cannot be assured of encountering a male

FIG. 4.14. The prairie hen males gather in a lek and display by inflating the large yellow air sacs on the neck and bowing with tail fan spread when the females approach. (After W. P. Pycraft, *Courtship of Animals*, 1914.)

as soon as she reaches the proper stage of estrus, in spite of powerful odor and seeking behavior. The natural pugnacity of the carnivore has thus been incorporated into the courtship of the animal and aids reproductive co-ordination by prompting ovulation.

The domestic rabbit is also known to remain in estrus without ovulation until released by the excitements of mating, although here mating is not accompanied by such rough-house play. This might be deemed a contradiction to the hypothesis formulated above, but perhaps it too can be understood in ecological terms. The female of the European wild rabbit is larger than the male and maintains a defended territory around her warren from which she excludes the smaller male. As might be expected under these circumstances, the male has perfected a courtship that is appropriately cautious but exciting to the female. He may walk in a stiff, high-stepping manner away from the female or around her, in either case with tail raised and rear quarters directed toward the female. Once the male excites and possibly dominates the female, he directs a jet of urine at her. Such "enurination" behavior is occasionally also seen in states of excitement provoked by being trapped, for instance. It is possible that the dependence of ovulation in the rabbit is a reflection of the difficulty in mating that results from the necessity of overcoming the territorial defense of the female. At any rate, in these few mammals, coordination must be regarded as a function of their courtship activities.

## Display as a General Phenomenon

If we think back at what has been said about display in relation to territorial defense, dominance patterning, and courtship, we recognize that these involve ostentatious "attention-calling" behaviors. All of these are of course related to reproduction to some extent; yet it is important to note that the functions of display extend beyond the strictly sexual relations of mates. In territorial display, the showiness is often directed not to the mate but to sexual rivals. This factor accounts for the prominent part played by aggression in the expressions of some displays. Because of the close interlocking of territoriality, dominance, and the reproductive activities, we cannot really identify many of these displays as being properly in one category or another; that is, reproductive behavior in social animals tends to become intermixed with these other two behavioral types so that characters which seem appropriate for one activity become deeply involved in the others. Thus defense of territory, rather than mating, seems to be the goal of "sexual" behavior in many cases. Some species of seals, for example, pay no attention to the females which may leave a territory; the male makes little attempt to herd them in yet he defends his territory zealously. In some gulls the attachment of the parent seems to be primarily to the territory, since this is ardently defended; whereas the eggs, nest, or mate can be changed around at certain stages without arousing any compensating behavior. From a behavioral standpoint, therefore, the sexual drive of an animal may be expressed more strongly in "irrelevant" activities than in those directly related to reproduction. It cannot, therefore, be properly said that defense of territory is a displacement of sexual activity or that dominance fighting by males is in any sense misdirected sexual motivation. According to the nature of the species, these activities may be of the essence of the expression of sexual motivation.

## Parental Care

Without attempting a systematic discussion of parental care in the animal world, we may consider here certain aspects which were not conveniently discussed earlier and which illustrate the adaptive nature of the variations to be found.

### PARENT-YOUNG INTERCHANGE

Among colonial insects, such as ants, termites, bees, and some wasps, it has been found that members of the colony actively exchange food by regurgitation. If dyed-sugar water is fed to some members of an ant colony, the color is seen throughout the colony within a few days. This exchange of food is called "trophallaxis." It plays an important part in the correlation of the activities of the insect colony. In termites, for example, it is believed that substances transmitted in the trophallactic pool determine the differentiation pattern of the young with respect to caste specialization. This trophallactic exchange also plays a role in parent-young relations (Michener and Michener, 1951). In ants, for example, the larvae produce fatty secretions for which the adults have a great appetite. The feeding of young by adults is thus an exchange phenomenon dependent upon the specific appetites, not upon young-parent recognition. This is illustrated by the success certain nest parasites, such as the Lomochusa beetle, achieve because they produce the sought-after secretions in even higher quantity than ant larvae. The adult ants feed them in preference to, and to the detriment of, their own larvae. The concept of a mutual exchange forming the behavioral basis for parent-young interrelations may be applied to many activities other than food exchange. In army ants, for example, the activities of the young in various stages of their development appear to provide the stimuli that regulate colony activities (Schneirla, 1938). Whether or not we use the word trophallaxis to describe the exchanges provided by stimuli other than food, the basic idea that the relation between parent and young is integrated by a mutual exchange rather than by a one-way giving is applicable to many other organisms.

An almost universal characteristic of mammalian maternal behavior is the eating of the placenta and the licking of the young by the mother. Even guinea pigs and many herbivores which normally do not eat meat at all eagerly consume the afterbirth. The licking of the young has certain obvious practical values—it cleans and dries them and clears their nose, mouth, eyes, and ears. The eating of the placenta, despite its high hormone content, is not definitely known to be of any practical importance for the mother. Only the higher primates as a group seem to lack this licking and eating behavior, although among domesticated herbivores (cows, for instance), the failure to eat the placenta is not uncommon, and there are several species of mammals, such as the wild pig and fur seal, which apparently do not eat the afterbirth (Hediger, 1955).

From the behavioral standpoint, the post-parturitional licking activities of the mother appear to be important in establishing the mother-young bond. It has been found in goats that if the young is removed immediately after birth and before the mother has licked it and is cleaned up and returned a few hours later to the mother, she will refuse to accept it. She no longer has the drive to lick the baby. As the young approaches and tries to nurse, she drives it away just as a normal goat drives away foreign young that try to nurse.

The ability of the mother to become associated with her own young seems to be limited to a short period, two or three hours in goats and sheep. Even slight interferences with the normal relationship during this critical time has far-reaching effects on the mother-young relationship. Experimental disruptions at this time, even if they do not prevent the baby from nursing but merely shorten the licking process, may result in a failure in the completeness of the mother's response to the young when they call from a distance (Blauvelt, 1955). The cleaning of the new-born chimpanzee is accomplished by finger manipulation by the mother. The behavior is akin to grooming and is visually guided and not dependent upon the chemical senses. As we shall see later, it also does not have as firm an instinctual basis as the licking behavior appears to have. In herd-living mammals the initial relationship between mother and offspring seems to be that of the mother recognizing her own young. The young soon learn to associate specifically with their own mothers presumably as a result of rejection by all other females. Thus, herds of many ungulates such as mountain sheep and deer have as their primary group the maternal family in which the young are associated with their mothers and these, in turn, with their own mothers as long as the latter survive. Since the mother retains a dominance and leadership relation to her young, the herd constitutes primarily a matriarchy with one or a few oldest females exerting social control (Darling, 1937; Murie, 1944).

A phenomenon analagous in many respects to what we have described above for mammals seems to take place in some birds during the establishment of parent-young relationship. Precocial birds, which follow their parent soon after hatching, have been found to become "imprinted" with the characteristics of the parent or surrogate parent during the first few hours after hatching. If a turkey or greylag goose is incubator-hatched and cared for by a human in its initial hours, it will thereafter show no fear of human beings but will instead react to them as it normally would to a parent. Greylag geese thus imprinted follow a person and will not follow adult geese. After the first few hours, the hatchlings lose this capacity for imprinting and will not form the association. Once imprinting is accomplished, it cannot be displaced from the imprinted object by other associations after this time. It should be noted that the process of imprinting is a young-to-parent relation, not a parent-to-young relation as we saw it develop in the ewe. The imprinting process shows a surprisingly long-range effect. For example, hatchling jackdaws imprinted to humans displayed courtship behavior only to humans and not to fellow jackdaws when they become sexually mature many months later (Lorenz, 1952). In birds that are hatched in a very immature condition and remain in the nest for a long period (altricial birds), the identification of young to parent, in so far as it occurs at all, is a more gradual process and takes place later.

The parent-young relationship is the primary social relationship in birds and mammals generally. It may remain in force with modifications throughout the life of the animal, as we mentioned above in the case of the red deer, and constitute the basis for herd organization. In mammals which live in permanent groups, this mother-young relationship is the basic factor in social organization. In birds the parent-young relationship usually disintegrates before the coming of winter. The migratory flocks which form then do not have a familial basis. Often older males, older females, and yearlings form separate flocks for migration.

In both birds and mammals the breakup of the original family relationship has been ascribed, at least in some instances, to positive disruptive behaviors rather than simple cessation of the bonds that held the individuals together. Thus when the female of many species of deer arrives at late pregnancy, she adopts a negative attitude to her yearlings and when they attempt to nurse rejects them with threats, butts, or even biting (Altmann, 1952). Mother bears actively take steps to break the associations between themselves and their cubs. In chickens, too, an active driving-off of the young is seen when the hen is ready to enter a new laying period. Where there are a few dominant males in the social groups, as in macaque monkeys, these males drive out the young males as they mature but do not eject the females. As a consequence, these monkey colonies generally have small bachelor groups which associate peripherally with them. The young males in this group are ever alert for an opportunity to invade the group in order to achieve a permanent place and gain access to the females.

## TEMPORAL LIMITATIONS

Another aspect of parental behavior that deserves particular comment is the timed, sequential co-ordination of the separate events. The fanning activity of the male stickleback, for example, is part of a timed, sequential behavior pattern. It begins only after a female has laid eggs in the nest and increases in intensity gradually and appropriately as the young develop. When they emerge from the nest, the fanning ceases abruptly. Similarly, the retrieving of young which have been removed from the nest constitutes one of the characteristic parental behaviors of mammals such as the rat. This behavioral response to misplaced young appears characteristically in the female late in pregnancy. It reaches a high peak during the first days after parturition and declines gradually so that by the time the young are ready for weaning on the twentieth day or so, retrieving behavior has practically disappeared. But, of course, it is in birds with their highly complex parental behaviors that we find this timing factor most prominent. A definite migration-restlessness appears in the appropriate season; territory-establishing behavior follows and lasts only a limited time. Song birds develop an interest in nest sites and nesting behavior which ceases at an appropriate time. Only then do the mates commonly show coitional behavior. This in turn may last only a few days and then disappear entirely from the repertory of the pair, and so on, through the entire sequence of their reproductive behaviors (Lehrman, 1961). We have previously emphasized the importance of courtship, territoriality, flock organization, and parent-young relations in maintaining the behavioral contacts and associations that permit necessary co-ordination. It must, however, be emphasized that the problems presented by the intricacy of these co-ordinations are among the most baffling in behavior study (Marshall, 1961). The results of the experimental analyses of these interactions will be examined in Chapters 5, 6, and 8.

## Sexual Dimorphism as a General Phenomenon

The sexual structures of animals are generally classified as primary and secondary. The primary include the sex glands, their ducts, and other accessories which serve the basic reproductive mechanisms characteristic of the taxonomic group. Since reproductive processes are fundamentally the same in all members of a group such as birds or placental mammals, the primary sex characteristics do not differ greatly between species in such groups. The secondary characteristics consist of those relatively superficial characteristics such as display feathers, horns, etc., which distinguish male from female of a species but which do not play a direct role in reproductive physiology. These are found to vary greatly from species to species and to be correlated with the detailed behavioral characteristics of the species. Sometimes another term, "epigamic," is used to refer to characteristics which, like secondary sex characteristics, play an indirect role in reproduction but which, unlike secondary sex characteristics, do not necessarily differ between male and female. The plumes of night herons or the red breast of robins might be cited as examples. These are alike in male and female but play a role in display behavior during reproduction. From the point of view of behavior, the important concept to be developed is that the secondary sex, or the epigamic, characteristics are closely related to behavior and often hold the clue to an understanding of the behavioral organization of a species.

One type of secondary sex characteristic may be described as physiological. It consists of a sensory-motor differentiation based on sex. This type is common among insects. The glow worm, for example, is the wingless, grub-like female of a firefly beetle. In many insects the male is differentiated as the motile, active, seeking type, whereas the female is the larger, food-storing specialist. This is analogous to the primary differentiation of the sperm and egg. This type of differentiation is readily understandable on physiological grounds since it contributes to the efficiency of the reproductive process. We would,

FIG. 4.15. The male (*right*) in the American bison illustrates the common characteristics of dimorphism in social-living mammals. The male is larger, especially in the fore-quarters. Head and horns are enlarged as part of the emphasis on aggressive potential. (After W. P. Pycraft, *Courtship of Animals,* 1914.)

therefore, expect to find it widespread. Yet it is most significant to note that among vertebrates this type of differentiation between the sexes is actually a rarity. An example is the deep-sea angler fish in which the female is large and fully developed, whereas the male is reduced to a tiny parasite.

A second type of secondary sex characteristic, which may be designated as aggressive potential, is the difference between male and female with regard to capacity for fighting (Fig. 4.15). This type is common among vertebrates. Most prominent among these dimorphisms are differences in size and strength. One of the extreme examples is seen in the seals and related marine carnivores. Elephant-seal males are as much as two and a half, and fur-seal males ten, times as large as their females (Bartholomew, 1952). Though this is extreme, a difference of 50 per cent or so is not at all rare among mammals.

In a majority of mammals, the male tends to be bigger and heavier than the female. Only exceptionally, as in the European rabbit, is the female the larger.

Aggressive potential in favor of the male often takes the form of weapons. Horns and antlers are in many instances differentiated between sexes. We are familiar with them in many species of deer. Teeth as weapons are also frequent secondary sex characteristics of mammalian males. We see this in the enlarged canine teeth in male baboons and, in extreme form, in the single large tooth of the narwal. In birds, examples of dimorphism in weapons are fewer, but the spurs of the rooster provide a good one.

A third category of secondary sex characteristics may be designated display characteristics. We have discussed many examples of these

FIG. 4.16. In the courtship of sexually dimorphic animals, the conspicuously colored male generally deports himself in such a way as to display his plumage patterns before the female. The snow bunting male, for example, faces away from the female and spreads his feathers stiffly, thereby exposing the conspicuous patterns of his back. He then runs quickly away from her for a short distance and repeats this performance again and again. (From N. Tinbergen, *Trans. Linn. Soc.*, 5 [1939].)

in the conspicuous plumage of many male birds (Fig. 4.16) and the color display of sexual skin in female primates. Perhaps some fancy furs among mammals deserve special mention. The male of many deer develops a thick coat of long and light-colored fur over the neck at rutting season. The lateral presentation of the "fur collar" is a prominent feature of display. The mane of the lion is probably to be regarded in the same category. Most of the better-known examples of sexual dimorphism among lower vertebrates also fall into this category. The brilliant colors of lizards and various aquarium fish, such as the fighting betta, immediately come to mind.

Several important points emerge from this brief consideration of secondary sex characteristics. It might be expected that natural selection would favor the physiological type of dimorphism, since each sex is thereby better equipped for its natural role. The female would be expected to be larger, since she must furnish the egg with its food supply or protect and nourish the young. The males would be expected to be, like sperm, small, numerous, and equipped with sensory-motor mechanisms that enabled them to seek out the larger and more sluggish females. As we have noted, such dimorphism is commonly found among invertebrates, particularly insects. Vertebrates, on the other hand, emphasize dimorphism of aggressive and display potential. In some ways this is the direct opposite to what might be expected.

The explanation of this paradox is to be found in the peculiarities of the position of the male in vertebrate reproductive patterns. Often the male contributes little more than the sperm transmitted to the female in the act of insemination. Yet to achieve insemination, he must compete with other males. Furthermore, since one male is physiologically capable of inseminating many females, a high survival rate for the males is of small consequence to the survival of the species. The selective processes to which such males are exposed thus favor characteristics which give success in competition with other males, with little regard for the deleterious effects which such characteristics have on the survival of the males themselves. Selection in females, on the other hand, is much more conservatively managed, for a high survival rate of the female is of primary importance, since care and provision for the young are dependent upon the female. Her more limited reproductive potential, moreover, makes her survival of great importance for the species' success. In this sense the males may be described as expendable in contrast to the females.

We can thus visualize that in the typical herd-living mammal, such as the deer, the primary determinant of biological success in a male is the capacity for maintaining a large harem in competition with other males. Characteristics of aggressive potential and display have been pushed by selection to the point of decreasing the chance of ultimate survival of the males bearing them. Of course, the need for individual survival cannot be entirely ignored, and there are limits beyond which the unphysiological specialization of the male for aggression or display cannot go. But it is clear that a characteristic that insures reproductive success may be favored even at the expense of individual survival. This is especially clearly shown in polygynous mammals. The existence of these selection pressures enables us to understand the general unphysiological specialization of males in many vertebrates and the well-known fact that, from a physiological view as measured by life span, resistance to disease, etc., the male is commonly inferior to the female of the same species.

It can be appreciated that the above reasoning applies particularly to mammals which live in groups and in which the male plays no role in parental care. It also applies to birds and other vertebrates under the same limitations. If we examine the instances of high development of display potential in birds, we see that the same principle can be invoked here. The birds which show the most elaborate development of conspicuous feathers are birds in which the male plays little or no role in caring for the young. The gallinaceous birds, lek-breeding birds, birds of paradise, etc., are all birds in which the parental functions are performed by the female to the practical exclusion of the male.

On the other hand, if we consider vertebrates in which the role of the male and female in care of the young is more nearly equal, such as the shore birds and most song birds, we find that the species is generally monomorphic, i.e., males and females are alike (Fig. 4.17). In these forms, we may find a moderate development of display potential in epigamic characteristics. This has its explanation in the value of the displays for mutual courtship and for territory defense on the part of both male and female. We do not, however, see such display potential carried to the point of interfering with individual survival by unduly exposing the bearers to predation.

FIG. 4.17. One form of nest-relief ceremony in the herring gull is for the bird returning to the nest to bring nesting material, as in courtship. Both parents in this monomorphic species share in care of the young. (After N. Tinbergen, *Herring Gull's World,* 1953.)

In exceptional instances, as in button quail and the phalaropes, where the behavioral role of the male and the female are reversed, it is not surprising to find that dimorphism also assumes a reverse pattern (Kendeigh, 1952). The female here is the showy-feathered member of the species (Fig. 4.18). Of course, in the vast realm of nature, where many other factors come into play, some exceptions occur, but it is an important general principle that the type of sexual dimorphism shown by a species correlates with the role of the sexes

FIG. 4.18. The male of the red-necked phalarope is smaller and less brightly plumaged than the female; he has considerably less red coloring on his throat, and the white stripe over the eye is inconspicuous. The female courts the male by rising up in the water, fluttering the wings, and calling loudly. Mating takes place in the water. The male builds the nest on the shore, and after the female lays the eggs, he assumes the entire burden of incubating. He pulls the tall grass over himself after settling on the nest for concealment. (Drawn from a photograph.)

in courtship and parental activities. In Chapter 10 we shall see the applicability of this principle to the understanding of social organization among vertebrates.

## REFERENCES

Altmann, M. 1952. Social behavior of elk, *Cervus canadensis* Nelsoni, in the Jackson Hole area of Wyoming. *Behaviour,* **4:** 116–43.

Armstrong, E. 1947. *Bird Display and Behaviour.* New York: Oxford University Press.

Aronson, L. R. 1957. Reproductive and parental behavior. *In:* M. Brown (ed.), *The Physiology of Fishes,* Vol. 2, pp. 271–304. New York: Academic Press.

BARTHOLOMEW, G. 1952. Reproductive and social behavior of the northern elephant seal. *Univ. Calif. Publ. Zool.,* **47:** 369–472.

BIRCH, H. G., and CLARK, G. 1946. Hormonal modification of social behavior. *Psychosom. Med.,* **8:** 320–31.

BLAUVELT, H. 1955. Dynamics of the mother-newborn relationship in goats. *In:* B. SCHAFFNER (ed.), *Group Processes.* New York: Josiah Macy, Jr. Foundation.

BOURLIERE, F. 1954. *The Natural History of Mammals.* New York: Alfred A. Knopf.

BRISTOWE, W. S. 1941. *Comity of Spiders.* London. The Ray Company.

BURTON, M. 1954. *Animal Courtship.* New York: Frederick A. Praeger.

CARPENTER, C. R. 1942. Sexual behavior of free-ranging rhesus monkeys (Macaca mulatta). *J. Comp. Phychol.,* **33:** 133–62.

CLARK, E., and ARONSON, L. R. 1951. Sexual behavior in the guppy, *Lebistes reticulatus* (Peters). *Zoologica,* **36:** 49–66.

COLLIAS, N. E., and JAHN, L. R. Social behavior and breeding success in Canada geese (*Branta canadensis*) confined under semi-natural conditions. *Auk,* **76:** 418–509.

DARLING, F. 1937. *A Herd of Red Deer.* London: Oxford University Press.

———. 1952. Social behavior and survival. *Auk,* **69:** 183–91.

EMLEN, J. T. 1955. The study of behavior in birds. *In:* A. WOLFSON (ed.), *Recent Studies in Avian Biology.* Urbana, Ill.: University of Illinois Press.

GREENBERG, B., and NOBLE, G. K. 1944. Social behavior of the American chameleon (*Anolis carolinensia* Voigt). *Physiol. Zool.,* **17:** 392–439.

HEDICER, H. 1955. *Psychology of Animals in Zoos and Circuses.* New York: Criterion Books.

HEINROTH, O., and HEINROTH, K. 1958. *The Birds.* Ann Arbor: University of Michigan Press.

HUXLEY, J. S. 1914. The courtships of the great crested grebe. *Proc. Zool. Soc. London,* **11:** 491–562.

IMANISHI, K. 1957. Social behavior in Japanese monkeys, Macaca fuscata. *Psychologia,* **1:** 47–54.

KENDEIGH, S. C. 1952. *Parental Care and Its Evolution in Birds.* Urbana, Ill.: University of Illinois Press.

LACK, D. 1953. *The Life of the Robin.* Rev. ed.; London: Penguin Books.

LEHRMAN, D. S. 1961. Hormonal regulation of parental behavior in birds and infrahuman mammals. *In:* W. C. YOUNG (ed.), *Sex and Internal Secretions.* Baltimore: Williams & Wilkins Co.

LORENZ, K. 1952. *King Solomon's Ring.* New York: Thomas Y. Crowell Co.

MARSHALL, A. 1961. Breeding seasons and migration. *In:* A. MARSHALL (ed.), *Biology and Comparative Physiology of Birds.* New York: Academic Press.

MICHENER, C., and MICHENER, M. 1951. *American Social Insects.* New York: D. Van Nostrand Co.

MURIE, A. 1944. *The Wolves of Mount McKinley.* ("Fauna of the National

Parks of the U.S.," No. 5.) Washington, D.C.: Superintendent of Documents.

NICE, M. M. 1937. Studies in the life history of the song sparrow, I. *Trans Linnaean Soc.* (N.Y.), **4:** 1–247.

NOBLE, G. K., and ARONSON, L. R. 1942. The sexual behavior of Anura, I. *Bull. Amer. Mus. Nat. Hist.*, **80:** 127–42.

NOBLE, R. 1945. *The Nature of the Beast.* New York: Doubleday Doran Co.

PYCRAFT, W. P. 1914. *The Courtship of Animals.* New York: Henry Holt & Co.

SCHNEIRLA, T. C. 1938. A theory of army-ant behavior based upon the analysis of activities in a representative species. *J. Comp. Psychol.*, **25:** 51–90.

STONOR, C. 1940. *Courtship and Display among Birds.* London: Country Life.

TAVOLGA, W. 1956. Visual, chemical and sound stimuli as cues in sex discriminatory behavior of the Gobiid fish, *Bathygobius sporator. Zoologica,* **41:** 49–64.

TINBERGEN, N. 1953. *The Herring Gull's World.* London: William Collins Sons & Co.

———. 1953. *Social Behaviour in Animals.* New York: John Wiley & Sons.

ZUCKERMAN, S. 1932. *Social Life of Monkeys and Apes.* New York: Harcourt, Brace & Co.

FRANK A. BEACH

# 5

# Biological Bases for Reproductive Behavior

## Introduction

This discussion will deal principally with evidence from studies on mammals, since most of the analytic data available apply to this group. Reproductive behavior will be considered as influenced simultaneously by three classes of variables: (*a*) neural, (*b*) hormonal (or biochemical generally), and (*c*) experiential. Before the discussion of these categories the general subject will be introduced by a specific example.

## Control of Mating Behavior in Female Mammals

At approximately seventy days of age the female rat displays sexually receptive behavior for the first time in her life. The pattern is a relatively simple one, which has been described in the preceding chapter and analyzed quantitatively in an earlier study (Kuehn and Beach, 1962). The estrus female approaches the male and then responds to his investigation by running away. Her gait consists of a jerky, hopping motion shown only while she is in heat. The male follows and mounts the female from the rear, clasping her sides with his forelegs. She responds by the assumption of lordosis, the position characteristic of, and limited to, the estrus female (see Chap. 4). One other response that may appear during estrus is a rapid shaking of the head in the lateral plane, a motion which produces violent vibrations of the ears.

At the time that these behavioral characteristics are in evidence, the female's vaginal lumen contains nucleated and cornified epithelial

Frank A. Beach is professor of psychology at the University of California, Berkeley.

cells, a condition not seen prior to the first heat period. Both the vaginal changes and the novel behavior are reactions to an alteration in hormonal condition. Several days prior to the onset of these symptoms, a number of egg cells in the female's ovaries began to mature. As this occurred the egg follicles became enlarged, and the thecal cells of the follicular wall began to secrete the hormone estrone. This hormone, acting upon the lining of the uterus and vagina and also upon critical circuits in the nervous system, prepared the female for the advent of sexual receptivity and induced cellular changes in the reproductive tract.

A few hours prior to the onset of behavioral estrus, a second ovarian hormone, progesterone, was released into the blood stream. Most of the progesterone produced by female mammals is secreted by the *corpus luteum,* but small amounts must be present before this time, as indicated in Chapter 2. They arise, presumably, from the ripe follicle. This conclusion is dictated by the finding that full behavioral estrus cannot be induced by injections of estrogenic hormone but depends upon estrogen-priming followed by administration of progesterone.

It remains only to be said that the female rat's pattern of sexual behavior is normal the first time it appears. No practice is needed to perfect it, and it remains unmodified throughout the reproductive life of the individual.

The initial appearance of mating responses obviously depends upon the capacity of the nervous system to respond to ovarian hormones, and it has been found that this capacity develops long before the secretory activity of the ovary begins to affect the sex ducts or behavior. Females only thirty days old will exhibit mating reactions indistinguishable from those of the sexually receptive adult if they are injected with the appropriate amounts of estrogen and progesterone. Evidently the neuromuscular mechanisms involved in mating reach maturity well in advance of the time that they normally are called into play.

Development of the ovarian follicle and the secretion of ovarian hormones are stimulated by two gonadotrophic hormones secreted by the anterior pituitary gland (see Chap. 2). This means that adult sexual reactions cannot occur until the ovary develops its responsiveness to pituitary secretions. Experiments have shown that such responsiveness is present in prepuberal females, and if exogenous gonadotrophins are administered to immature female rats, the adult pattern of estrus behavior promptly appears.

It would apear that the key factor controlling the timing of the first estrus rests upon the production of gonadotrophic hormones by the anterior pituitary gland, since the nervous mechanisms for mating and the capacity of the ovaries to secrete estrogen and progesterone are fully developed before puberty. It has been shown, however, that when the pituitary glands of immature females are transplanted to the *sella turcica* of immature females, the transplanted glands release gonadotrophic hormones capable of stimulating ovarian growth and endocrine function. The conclusion is that the immature pituitary can secrete gonadotrophins but does not normally do so unless stimulated by the mature brain. Hormone-release is triggered by events within the nervous system which are transmitted by way of the hypothalamic-hypophyseal portal blood vessels (see Chap. 2). This hypothesis is strengthened by experiments involving the transplantation of pituitaries from male to female rats. The female's own pituitary is totally removed and replaced by the same gland removed from a male. When such grafts become sufficiently vascularized, the experimental females show normal estrus cycles. This is taken to signify that, under normal conditions, the cyclic release of gonadotrophic hormones by the pituitary in the female is under neural control through the hypothalamus.

We have mentioned the role of neural and endocrine factors in control of the female's sexual behavior. In the case of the rat, individual experience seems to exert little or no influence upon the mating pattern. In higher animals, however, and particularly in anthropoid apes, this is not the case. Even though the hormonal agents are present in sufficient amounts, and even though the neural apparatus is sufficiently well organized to mediate all of the individual elements in the female's coital pattern, successful mating cannot be achieved in the absence of previous experience and practice (Harlow, 1962).

With this brief introduction to the three classes of factors affecting reproductive activity, we may now turn to a more detailed discussion of the evidence pertaining to each category.

## Neural Factors in Sexual Behavior

At various times in the history of neurology attempts have been made to locate a "sexual center" in the brain. The German anatomist and phrenologist, Franz Joseph Gall, reached the conclusion that the "amatory instinct" must have its seat in the cerebellum, because the skull overlying this area was unusually warm in a hysterical widow of his acquaintance. Other speculations implicated the hippocampus, because that structure was thought to be intimately associated with the sense of smell, and olfactory stimuli were assumed to be of paramount importance in sexual arousal. More rigorous methods of accumulating evidence on this score have involved studies of neuropathologies, correlations of neuroanatomy with the innervation of sexual structures, and the experimental modification or recording of brain function in association with observations upon behavior.

### THE CEREBRAL CORTEX

At one time it was a common practice to make a sharp distinction between behavior patterns considered to be "instinctive" and those which were acquired by the individual. It was sometimes postulated that inherited behavior is mediated by neural tissue lying below the cerebral cortex while the more recently evolved neocortex is concerned with the learning of new habits. According to this point of view, one might predict that the sexual and parental behavior of "lower" mammals might be mediated subcortically and occur independently of the neocortex.

Evidence in favor of such an hypothesis has been reported for the male rabbit. Extensive injury to the cerebral cortex of this animal does not eliminate mating behavior (Brooks, 1937). On the other hand, systematic studies of the effects of cortical injury to male rats clearly show that large cortical lesions abolish the tendency to copulate with receptive females (Beach, 1940). Smaller lesions may spare the ability to copulate but nevertheless reduce the probability of mating (Fig. 5.1). The most reasonable conclusion is that the neocortex is not essential to the execution of the motor pattern of coition but does contribute to the occurrence of sexual arousal. Males with extensive cortical loss may be physically capable of copulating but fail to display mating behavior because they never become sufficiently aroused.

A striking finding, which has been verified repeatedly, is that female rats and rabbits are capable of fertile union with normal males after total removal of the cerebral cortex. The rat's receptive pattern is somewhat disorganized after such an operation, but it remains suffi-

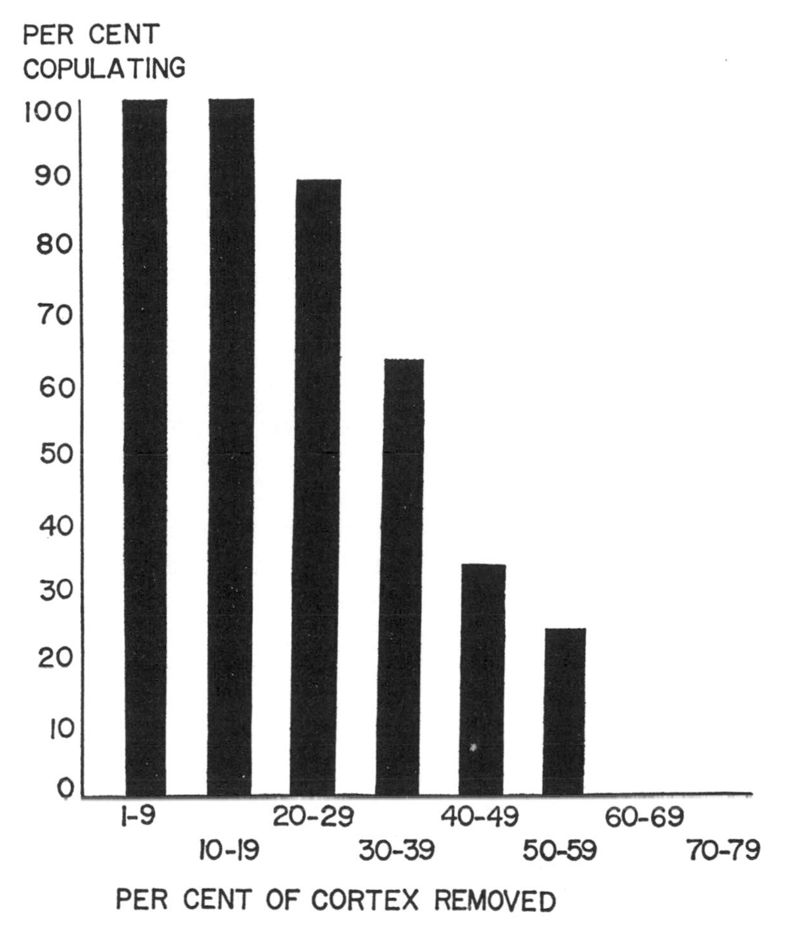


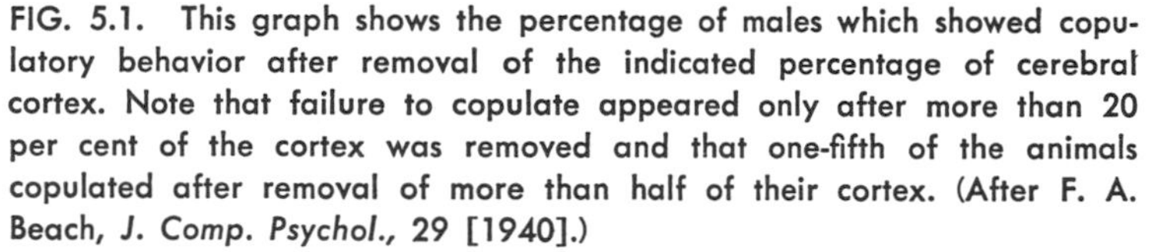

FIG. 5.1. This graph shows the percentage of males which showed copulatory behavior after removal of the indicated percentage of cerebral cortex. Note that failure to copulate appeared only after more than 20 per cent of the cortex was removed and that one-fifth of the animals copulated after removal of more than half of their cortex. (After F. A. Beach, *J. Comp. Psychol.*, 29 [1940].)

ciently normal to permit effective copulation (Beach, 1944; Brooks, 1937).

One nineteenth-century experiment showed that male dogs deprived of the forebrain display no tendency to copulate with the estrous bitch, although females of this species will receive the male after the same operation (Goltz, 1892). More fully documented investigations have shown that decorticated female cats can mate successfully, whereas males lacking a cortex are unable to do so (Beach, Zitrin, and Jaynes, 1956; Bard, 1936).

Experimental destruction of limited amounts of the neocortex has different effects upon the male cat's sexual performance depending upon the locus of the injury. Unilateral decortication does not prevent normal mating (Beach, Zitrin, and Jaynes, 1956). Bilateral removal of the occipital cortex including the visual areas causes blindness and therefore prevents the male from finding the female on the basis of visual cues (Zitrin, Jaynes, and Beach, 1956). Once physical contact is established, however, a male with this type of brain loss is perfectly capable of normal copulation. Male cats with extensive bilateral injury to the frontal cortex show gross abnormalities of sensorimotor adjustment. They display unequivocal signs of high sexual excitement in the presence of the estrus female, but they are generally unable to make the bodily adjustments necessary for successful copulatory union (Beach, Zitrin, and Jaynes, 1955).

These findings are consonant with the well-known fact that many aspects of sensory and motor function are more heavily dependent upon cortical control in carnivores than they are in rodents. It is not that the control of sexual functions per se has been transferred to the cortex, but that all kinds of behavior involving delicate co-ordination of sensory input and motor output have come to involve more of a cortical component.

This process of "corticalization" of function has been extended even further in the primates. Although direct evidence is lacking, it is reasonable to suppose, therefore, that the sexual activities of monkeys and apes would be even more seriously curtailed by appropriately placed cortical lesions.

### THE AMYGDALA AND ASSOCIATED LIMBIC STRUCTURES

In 1939 Klüver and Bucy reported complex alterations in the behavior of monkeys which had been subjected to temporal-lobe lesions. These included, among other things, a marked increase in sexual behavior, specifically in the frequency of masturbation. Since the lesions also involved the underlying amygdaloid nucleus, it was subsequently suggested that the "hypersexuality" following the operation was due to invasion of this portion of the limbic system. More recent workers have described "hypersexuality" in male cats following direct destruction of the amygdala, and in some cases, after destruction of portions of the rhinencephalic cortex (i.e., the limbic system—see Chap. 2 and Schreiner and Kling, 1953). The behavioral symptoms included persistent attempts to copulate with other male cats, with chickens, with monkeys, and even with inanimate objects such as toy animals.

Schreiner and Kling's report as well as that of Klüver and Bucy are difficult to interpret because in neither case is there any account of the preoperative sexual behavior of the experimental animals. There is no base line against which to measure the supposed changes subsequent to brain injury. In fact, the critical reader is left with the distinct impression that the postoperative sexual activity was discovered quite by accident and the absence of any preoperative tests was due to a failure to anticipate this phenomenon.

A later experiment by Green, Clemente and deGroot (1957) yielded somewhat different results. In the first place these workers observed their cats both before and after brain operation and noted that normal males may exhibit several sexual responses which Schreiner and Kling had attributed to amygdalectomy. These included tandem mounting, "homosexual" mounting, and attempts to copulate with anestrus females and small kittens. These types of behavior occurred only within a male's "home territory." No mounting of inanimate objects was shown by untreated males, but one intact cat implanted with testosterone repeatedly exhibited copulatory responses to a teddy bear.

Green and his co-workers placed lesions in the pyriform cortex and amygdala of sixty males and noted "hypersexuality" in twenty of these animals. In nineteen of the twenty cats the brain injury involved the pyriform cortex and in fifteen the amygdala was heavily invaded. Other animals with lesions restricted to the amygdala did not exhibit unusual sexual behavior. It was therefore concluded that such sexual changes as did occur in this study, and also in the experiment by Schreiner and Kling, implicated the pyriform cortex rather than the amygdala.

Unfortunately it is impossible to make pre- and postoperative comparisons on an individual basis, and one must be content with a general description. It is stated that the most pronounced changes following pyriform-cortex injury was the abolition of the "territorial effect." Operated males attempted copulation with a variety of stimulus objects, regardless of where they were encountered. One or more of the operates mounted a rat, guinea pig, anaesthetized cat, toy animal, and even the shoe of the experimenter.

Although it represents a distinct improvement on the work of Klüver and Bucy as well as that of Schreiner and Kling, the report of Green, Clemente and deGroot would have been greatly strengthened by the administration of standardized behavioral tests to all animals before and after operation. The importance of territoriality was not recognized until after a number of the animals had been operated and tested for sexual behavior. The experimenters were apparently unfamiliar with the range of variation which occurs in the sexual behavior of unoperated cats. Thus, although never observed before brain operation, sexual responses directed toward inanimate objects have been described by other workers. Rosenblatt and Aronson (1958) report the case of one male cat which developed the habit of holding its food dish on edge and rubbing the penis against it while maintaining the normal copulatory posture. This behavior, which often continued for an hour or more, often terminated with an "orgastic" response. Michael (1961) describes masturbatory behavior resulting in ejaculation. In this case the stimulus object was a toy rabbit.

Any reports of "hypersexuality" in brain-operated cats should be compared with Hagamen and Zitmann's description of similar behavior in unoperated animals (1959). Michael (1961) has made the same point we are stressing here:

> Activity shown by male cats towards anoestrous females, kittens, other male cats, inanimate objects and alien species has been used . . . as a criterion of abnormal hypersexuality. The observation of these behaviour patterns following lesions in, or destruction of, the amygdala and pyriform cortex has been used to implicate these structures, and the temporal lobe, very importantly in the regulation of sexual behaviour. The data presented here shows clearly that such behaviour, with the exception of the mounting of alien species, can occur either spontaneously or be produced (as a conditioning effect) by simple manipulation of the environmental situation. No observations were made with alien species and no comment is offered. Though such phenomena as multiple mounting seem at first to be very bizarre, the expression of identical behaviour, in animals which have not been subjected to any operative interference, suggests a need for caution in the interpretation of results (Michael, 1961, p. 20).

The original report by Klüver and Bucy attracted so much attention that a description of "hypersexuality" in brain-operated humans was probably inevitable. Terzian and Ore (1955) filled this gap with their account of the behavior of a nineteen-year-old male epileptic after bilateral removal of the temporal lobes. The symptoms mentioned consisted of exhibiting the genitals, frequent masturbation, and verbal solicitation of homosexual relations. No heterosexual interest was noted, and there is no mention of the patient's sexual habits prior to operation.

### THE HYPOTHALAMUS

It has long been known that various hypothalamic nuclei are closely associated with reproductive functions. This is not surprising in view of the intimate relationship between the hypothalamus and the anterior pituitary gland as discussed in Chapter 2. Certain types of hypothalamic lesions have been shown to disrupt the gonadotrophic functions of the pituitary and thus to create a condition of functional castration, which, in turn, is reflected in the loss of mating behavior.

Other hypothalamic areas appear to contribute more directly to the mediation of mating behavior. For example, lesions at the level of the posterior border of the optic chiasma, one millimeter above the ventral surface of the brain, permanently eliminate sexually receptive behavior in female guinea pigs, and this deficit is not repaired by administration of very large amounts of estrogen and progesterone (Brookhart, Dey, and Ranson, 1940).

Male rats with lesions in the premammillary region of the ventral hypothalamus show a progressive loss of sexual responsiveness similar to that which follows castration. Sexual responsiveness is restored to normal by androgen treatment. Injury to the posterior ventromedial tuberal area greatly reduces the occurrence of mating responses, and this effect is not reversed by androgen treatment (Rogers, 1954). Judging from the scant evidence presently available, the hypothalamus appears to serve a dual function with respect to sexual behavior. It regulates the gonadotrophic functions of the anterior pituitary and thus controls the secretion of gonadal hormones; and it also contributes, in a manner not yet understood, to the neural mediation of mating activities.

It has been reported (Miller, 1957) that stimulation of the hypothalamus by way of chronically implanted electrodes can evoke sexual ejaculation in male rats. This effect is as yet not reliable and has proven difficult to repeat with any regularity. In another article it has been stated that "sexual effects" can be obtained in the rat by stimulating the septal region (Olds, 1958). Actually, the published findings bear no direct relation to sexual behavior. They show that males will perform an instrumental act which leads to septal stimulation. The frequency with which the act is performed is markedly reduced by castration and subsequently increased by the administration of androgen. These data, while intrinsically interesting and potentially significant, have no immediately apparent bearing on possible relations between septal mechanisms and sexual behavior.

A study of the male squirrel monkey has revealed that penile erection follows electrical stimulation of several different brain areas, including several subdivisions of the limbic system (MacLean and Ploog, 1962). Of particular interest is the added observation that the hippocampus modifies the excitability of structures involved in penile erection. Furthermore, some of the stimulations which result in erection are commonly associated with afterdischarges in the hippocampus, and during the afterdischarge "erections may become throbbing in character and reach maximum size." The possibility of hippocampal involvement in sexual climax inevitably suggests itself.

## Hormonal Factors in Mating Behavior

We turn now to a consideration of certain biochemical factors which are involved in sexual and other reproductive activities. For the most part, we shall be concerned with secretory products of the reproductive glands and the hypophysis. It is known that there are differences in hormonal functions between different classes and orders of vertebrates. Because of limitations both in space and available evidence, we shall here examine primarily the relations between hormones and behavior as they are known to exist in mammals.

### SOURCE OF EVIDENCE

Evidence concerning the relationship between hormones and reproductive behavior comes primarily from four different sources. Since certain endocrine glands do not become functional until after a certain age, it is often possible to trace correlations between the advent of maximal secretory function in a given gland and the appearance of a new kind of behavior. Correlational evidence of this sort is, of course, not proof of a cause-and-effect relationship. Nevertheless such correlations often give rise to interpretations about cause and effect which can then be checked experimentally. Other developmental changes in the hormone system are related to the effects of aging. Some glands either cease to function or undergo functional modifications with advancing years, and in such cases it is sometimes possible to describe parallel changes in behavior. Observations of this type often yield presumptive evidence concerning the behavioral effects of the hormones in question.

Some glands appear to secrete their endocrine products at a fairly steady rate throughout life. Others are cyclic in function, and may even be totally inactive for long periods of time (seasonal activity—see Chap. 2). When one is dealing with a gland whose products increase and then decrease at regular intervals, it often proves possible to identify concomitant changes in the individual's behavior which can be tentatively ascribed to the hormonal rhythms.

Important insight into the behavioral correlates of hormonal activity has often been provided by the study of pathologic conditions. It sometimes occurs that a given gland fails to function or functions at an abnormally low level (Heller, Nelson, and Roth, 1943). In other cases a gland may be chronically hyperactive, secreting more than the normal amount of its hormonal products. In either case it is frequently possible to discover behavioral abnormalities which parallel the endocrine pathology. From such correlations important deductions have been made concerning the normal behavioral effects of various endocrine products.

The fourth method—and the one with the most potential for obtaining reliable evidence—involves the application of the experimental method to problems in this area. This includes the possibility of stimulating secretory function in glands which would normally be inactive. It includes studying the effects of removal of a gland and consequent withdrawal of its hormone, or hormones. In the past, a number of experiments have been conducted in which glands have been removed from one individual and transplanted to a new host or a new site in the same individual. Finally, there have been many experiments in which extracts of a hormone-producing organ were administered and the behavioral effects of the treatment observed. In the case of some hormones it is now feasible to employ synthetically created chemical agents which are identical with, or simulate the effects of, the natural product.

### EVIDENCE FROM ONTOGENY

Study of developmental schedules, involving correlations between the onset of hormonal activity in a given gland and a related behavioral change, impresses the investigator with the fact that there are marked differences between species and, in some cases, even between males and females of the same species. This makes it difficult to summarize briefly the conclusions drawn from this sort of correlational study. Nevertheless it will be worth while to survey at least the high points of our knowledge in this area and to indicate in general terms the sorts of relationships that are known to obtain.

The male gonad (testis) begins to secrete its hormone (testosterone) during fetal life. The level of secretory activity is low at this time and remains so until approximately the age of puberty. Shortly before puberty, the amount of testosterone secreted by the male gonads rises rather sharply. In the human male there is a fairly steady increase in androgen level from approximately ten to fourteen years of age; and from fourteen to sixteen years the concentration of androgenic hormones increases abruptly in most individuals (Dorfman, 1948). There is reason to believe that the secretion of male hormone decreases in men at advanced ages, but the capacity of the testes to produce testosterone seems never to be lost completely.

Ovarian estrogen is absent or present in very low amounts in the human female until the onset of puberty, which is usually associated with the beginning of menstruation. Pronounced increases in estrogenic levels begin at approximately eleven years of age in some females; estrogen reaches high levels by fourteen years.

It is important to note that although the length of time necessary to attain reproductive maturity varies tremendously from species to species, the general chronological pattern of the onset of secretory activity in the gonads is fairly common to all species within the class Mammalia. This means that marked interspecific differences in mating behavior can probably not be explained in terms of endocrinological differences.

In a number of species the immature male shows certain elements of the adult mating pattern well in advance of the time that the testes exhibit the prepuberal rise in secretory activity. For example, male rats thirty to forty days of age engage in a great deal of playful chasing, wrestling, and mock fighting. In the course of a bout of such high-level activity, a male in pursuit of another young animal will sometimes grip the sides of the other individual and mount momentarily in a manner reminiscent of the sexual mount shown by the mature copulator. This behavior is always fragmentary and does not give the impression of being associated with a high level of sexual excitement. It might be thought that this adumbration of the adult mating response is due to the presence of very small amounts of testicular hormone. This clearly is not the case, because the same activity occurs in immature males which had been castrated on the day of birth (Beach and Holz, 1946).

In contrast to the prepuberal male, the sexually immature female rat does not display any of the behavioral elements characterizing the adult female in estrus. It has been noted above that such behavior can easily be induced in the preadolescent rat by the administration of estrogen and progesterone; therefore the absence of sexually receptive responses prior to puberty is probably referable to the absence of any appreciable amounts of ovarian hormones.

The foregoing description can be applied equally well to the development of sexual behavior in the Canidae. The manner in which the puberal rise in gonadal hormone affects the behavior of male dogs has been dramatically illustrated by Tinbergen (1951) in his account of the Eskimo dogs of East Greenland.

In each Eskimo village adult male dogs form packs, and each pack has its own territory which it defends against the members of all other packs. Adult males clearly recognize the limits of their own pack territory as well as those of their neighbors. Immature males seem unable to learn the territorial boundaries and as a consequence are continually wandering in and out of the property of other packs, with the result that they are frequently punished.

Tinbergen observed that the onset of sexual maturity was accompanied by three marked changes in the male dog's behavior. The act of urination was accompanied by elevation of one rear leg and by the direction of the urine stream upon some conspicuous object in the environment. At the same time there appeared a sudden increase in the male's interest in estrus females and in the tendency to show sexual responsiveness to them. Finally, the existence of territorial boundaries seemed suddenly to take on significance. Tinbergen observed these three changes taking place in one individual within a single week.

In the female dog, as in the female rodent, adult sexual reactions are lacking until the occurrence of the first estrus period. During her initial heat or "season" the bitch shows all of the responses characteristic of the sexually experienced female and will mate readily with an active male.

Among the ungulates, young males frequently exhibit mounting activity some time in advance of the achievement of reproductive maturity (Hooker, 1944). This behavior is not executed with any degree of forcefulness, and it usually appears within a context of various "playful" responses. Females of the same species do not exhibit sexually receptive activities until the ovaries go through the first cycle of estrogen secretion. At a comparable point in development, the mounting responses of the young male become more frequent and forceful and are directed specifically to females in estrus.

It seems clear thus far that for both male and female mammals the puberal rise in gonadal hormones is reflected in easily observed changes in behavior and that these changes include the appearance of the full copulatory pattern. This generalization applies to primates as well. The stage of puberty in young monkeys and apes is marked by a recognizable upsurge in erotic responsiveness and in the capacity for full coital activity. However, it appears that primates differ from "lower" mammals to some extent in the amount of sexual play which occurs before adolescence. Full copulation with orgasm has not been described, but exploratory attempts at heterosexual mating have been observed in young chimpanzees. In such circumstances the male and female execute most of the elements of the adult mating pattern but do so without any signs of high arousal or excitement (Bingham, 1928).

It is reported (Kinsey *et al.*, 1953) that human infants of either sex are capable of responding to genital stimulation with the display of a pattern which in many respects resembles the adult orgasm. It is also known that children in many, if not all, societies indulge in a variety of forms of sexual play well before the beginnings of adolescence (Ford and Beach, 1951). In as much as the endocrinological picture in the human species does not vary in any marked way from that seen in other mammals, behavioral differences during the prepuberal stages must be referred to some non-endocrinological cause, or causes. Nevertheless, it seems quite evident that the human capacity for sexual response and copulatory performance undergoes a marked increase during the early teen ages, and this may properly be ascribed to the puberal rise in the level of gonadal hormones.

### EVIDENCE FROM SEASONAL CYCLES

For the sake of completeness it is important to note that the annual appearance of mating behavior in many mammalian species is accompanied by, and dependent upon, a seasonal increase in gonadal activity. The relevant evidence is reviewed in Chapter 4 and need not

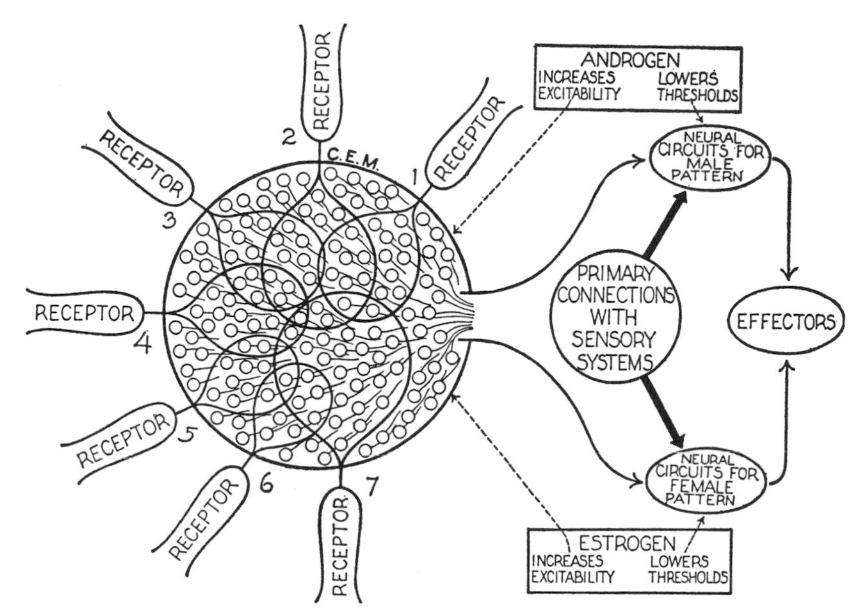


**FIG. 5.2. Relationship between the postulated central excitatory mechanism, the different sensory receptors, and the neural mechanisms for sexual motor patterns in rats. (After F. A. Beach, *Psychosom. Med.*, 4 [1942].)**

be repeated here. Suffice it to point out that most undomesticated species are capable of fertile union only during the mating season and that only during this relatively brief period are ripe eggs and sperm produced, and ovarian and testicular hormones secreted, in maximal concentrations.

## EVIDENCE FROM SHORTER CYCLES

In many domesticated species of mammals and in most primates, the adult male is sexually potent at all times, excepting those periods which follow sexual exhaustion resulting from unlimited opportunity for copulation. This is not true of the females of any mammalian species about which we have any detailed information.

Some females belonging to the seasonally breeding type of species are monestrus (see Chap. 4). Females of the type under discussion here are polyestrus. They may come into heat at regular intervals throughout the year, or they may have a succession of estrus cycles followed by a longer period of non-receptivity. An excellent and well-studied example of the polyestrus type of mammal is the domesticated rat. Females of this species ordinarily come into heat at intervals of four and a half to five days. For several hours they are maximally receptive to the male. At this time the vaginal smear is characterized by the presence of nucleated and cornified epithelium, and the level of estrogenic hormone is very high. As the female passes out of the receptive condition, the vaginal picture changes and the estrogen level falls. This sort of a relationship argues strongly for the thesis that the condition of behavioral receptivity in this species is closely related to the level of the critical ovarian hormones.

Longer but comparable cycles are seen in the guinea pig and in other laboratory and domestic farm animals, including the sow. In all of these instances the correlation between ovarian condition and behavioral receptivity is marked and suggests a causal relationship.

Among the primates the female's ovarian cycle is nearly as precise and seemingly automatic as it is in the infraprimate mammals. There is also a very obvious cycle of erotic responsiveness in non-human species of primates (Young and Orbison, 1944). The female monkey or ape—particularly the latter—may, however, under special conditions, exhibit sexual receptivity of a low order at times when she is not physiologically in estrus. This may be of considerable importance to our understanding of human sexuality. It would be easy to overestimate the frequency and completeness of sexual response in diestrus chimpanzees and monkeys, and the fact is that mating behavior which occurs when the female is not fertile is plainly distinguishable from that which occurs when ovulation is imminent. Nevertheless, the fact that it occurs at all is of considerable importance, in as much as its counterpart in infraprimate species has yet to be reported.

## EVIDENCE FROM EXPERIMENTAL STUDIES

As has been indicated earlier, the most compelling evidence indicating the importance of hormonal factors as they influence sexual behavior is derived from studies employing the experimental approach. It has been noted that the administration of pituitary gonadotrophin is capable of eliciting precocious sexual activity in female rodents (Smith and Engle, 1927); it may be added that the same is true of males of the same species (Steinach and Kun, 1928). This means that by inducing an acceleration of secretory activity in the ovaries and testes of lower mammals, one can produce a premature appearance of the adult mating pattern. This strongly suggests a heavy dependence of such behavior upon hormones secreted by the sex glands. This conclusion is further substantiated by observations to the effect that the same result is obtainable if the experimenter injects or implants the immature male or female with gonadal hormones (Beach, 1942).

The most obvious and direct approach to the experimental study of relationships between gonadal hormones and mating behavior involves the removal and replacement of sexual secretions. The results of such experimentation in infraprimate mammals are reasonably clear-cut and well established. It seems beyond question that surgical loss of the reproductive glands (and consequently of their endocrine products) before puberty prevents the development of normal adult mating behavior in both male and female rodents. Castration performed on the day of birth eliminates the normal adolescent upsurge of erotic responsiveness and copulatory ability in male rats. Nevertheless, animals subjected to this operation at such an early age exhibit unmistakable signs of a marked increase in sexual excitabliity if they are injected with androgenic hormones in adulthood (Beach and Holz, 1946). In the case of the female rat, even congenital absence of ovaries, uteri, and vagina does not prevent the display of normal sexually receptive responses in adulthood when such aberrant individuals are injected with the appropriate combination of ovarian hormones (Beach, 1945).

The effects of removing the reproductive glands in adult animals vary according to the species involved. The evidence pertaining to rodents plainly indicates that castration during adulthood eliminates sexual activity in both sexes and this change is reversed by administration of the appropriate gonadal hormones. Conclusions are less simple when interest centers upon the Carnivora. It seems well established that ovariectomy promptly and permanently abolishes sexual receptivity in female dogs and cats; but as far as the male is concerned, results of castration are less easily summarized, and copulatory activity may persist for months or even years after removal of the gonads.

Rosenblatt and Aronson (1958) found that male cats which were castrated in adulthood before they had acquired heterosexual experience showed relatively little sexual activity postoperatively. Some males were able to achieve intromission in tests conducted shortly after castration, but this ability was lost within a few weeks. In contrast, some cats with extensive preoperative mating experience continued showing the full copulatory pattern for from eight months to three and a half years. The frequency of intromissions per test declined in all cases, but the ability of sexually-experienced castrates to execute the normal mating response was clearly demonstrated.

Evidence presently being prepared for publication in the writer's laboratory shows that the effects of castration on mating in male dogs vary with the age at operation and the amount of preoperative sexual experience. Prepuberal castrates never copulate successfully even

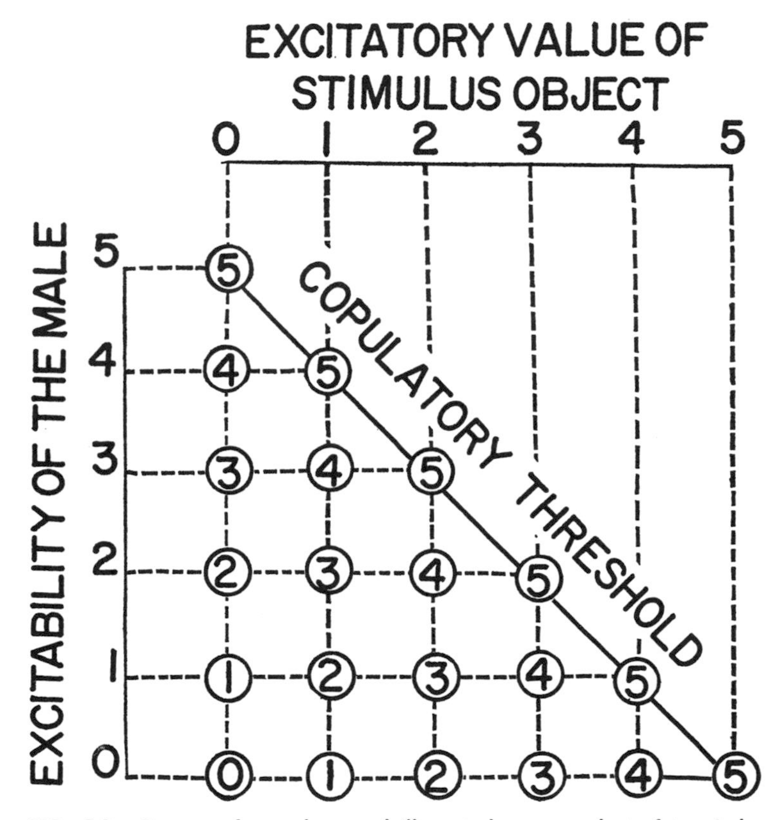


FIG. 5.3. Degree of sexual arousal illustrated as a product of two independent variables. (After F. A. Beach, *Psychosom. Med.*, 4 [1942].)

when they are treated with testosterone. Males castrated in adulthood without prior heterosexual experience do not mate without treatment but can do so after a series of androgen injections. The mating behavior disappears when hormone treatment is discontinued. Males castrated in adulthood after they have acquired experience with receptive females may maintain a high level of sexual performance for several years after operation. Some individuals show no decline in potency for at least three or four years. Others exhibit partial or complete decline in mating ability but can be restored to normal by androgen treatment.

There have been no systematic studies of the sexual behavior of gonadectomized monkeys or apes. The scattered evidence available (Beach, 1947) indicates that overiectomized females are generally unreceptive, although they may occasionally permit the male to copulate. Estrogen administration induces full behavioral estrus.

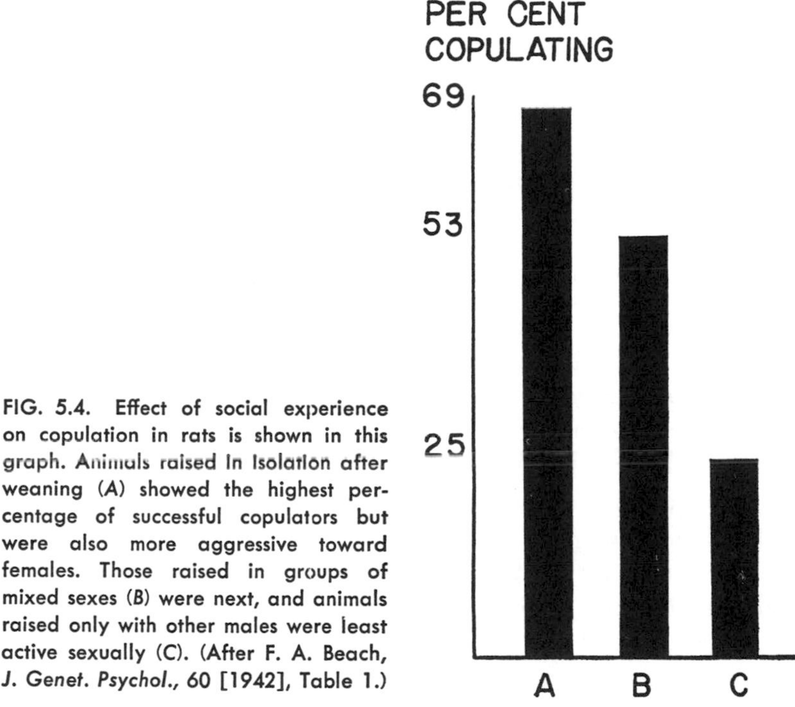


FIG. 5.4. Effect of social experience on copulation in rats is shown in this graph. Animals raised in isolation after weaning (*A*) showed the highest percentage of successful copulators but were also more aggressive toward females. Those raised in groups of mixed sexes (*B*) were next, and animals raised only with other males were least active sexually (*C*). (After F. A. Beach, *J. Genet. Psychol.*, 60 [1942], Table 1.)

Castrated males seem to retain their potency, although the capacity for ejaculation and orgasm is lost.

Clinical reports suggest that the human female is capable of full erotic responsiveness in the absence of ovarian hormones. This conclusion is based upon studies of the effects of natural and surgically induced menopause (Filler and Drezner, 1944). Castration of adult men does not necessarily eliminate potency, although the medical records are rarely completely satisfactory, and in some, great emphasis is placed upon so-called psychic factors (Kinsey, Pomeroy, and Martin, 1948).

## Experiential Factors

Although for the time being the factors discussed in this final section are treated as though they were extra-physiologic in nature, we take it as a matter of faith that the effects of individual experience and learning eventually will be understood in terms of changes in the nervous system. There are two areas in which individual experience could conceivably modify sexual activity. The first is on what might simply be called the "stimulus side." That is to say, individual experience might modify the stimulus complex capable of evoking sexual arousal and performance. Alternatively, practice or habit could theoretically modify the motor acts through which sexual excitement expresses itself. These possibilities have been investigated to a limited extent, and the one clear finding is that there are marked differences between species and even between sexes within the same species.

### MODIFICATION OF THE STIMULUS PATTERN

It seems reasonably clear that for both male and female mammals, at least those which belong to infrahuman species, the initial sexual arousal is ordinarily limited to a stimulus pattern presented by an individual of the opposite sex in reproductive condition. At the same time it should be noted that, as in any stimulus-identification situation, description of the "adequate" stimulus involves more than one variable. That which is "adequate" for one individual may be less than or more than "adequate" for another. In fact the adequate stimulus can only be defined in terms of the excitability or arousability of the responding individual, and there appears to be an inverse relationship between the sexual excitability of the test animal and the specificity of the minimally adequate stimulus pattern.

In spite of these qualifications, the generalization still obtains and indicates that in the vast majority of instances sexually naïve males of a given species are aroused to the point of copulation only by stimuli presented by sexually receptive females of their own kind, and the converse applies to the females. As far as the motor pattern of courtship and copulation is concerned, the statement is even more cogent. Modifications of this pattern on the basis of individual experience and habit-formation are almost completely unknown in males and females of infraprimate species.

This does not mean that individual experience is unimportant or uninfluential in the sexual behavior of "lower" mammals. Although the initial arousal is primarily limited to the stimuli provided by a biologically adequate partner, this situation may be modified rather rapidly as a result of individual experience. For example, it has been shown that male rats or males of some infraprimate mammalian species tend to become sexually aroused when they are placed in environmental surroundings in which they have previously experienced complete sexual activity including ejaculation. Thus, male mammals which have been repeatedly bred in a particular crate, pen, or cage are very likely to attempt sexual congress with non-receptive females or even with males presented under the same circumstances at a subsequent time (Beach, 1947).

It is also quite clear that a similar kind of conditioning can result in sexual "inhibition" in males at this same phyletic level. Animals which are consistently punished for attempting sexual congress in a given environmental situation may subsequently refuse to mate with receptive females in that particular surrounding (Beach *et al.*, 1956). Furthermore, when mating does occur in an "anxiety-producing" environment, various abnormalities such as premature ejaculation may appear (Beach and Fowler, 1959).

It is less certain that a comparable kind of conditioning occurs in females of the same species. The evidence is not systematic but only suggestive. What it suggests is that females at this (infraprimate) level are much more difficult to condition either for or against sexual activity by previous experience.

### MODIFICATIONS OF THE MOTOR PATTERN RESULTING FROM EXPERIENCE

Given the fact that the motor pattern of expression of sexual excitement is well developed and biologically effective in both males and females of the lower mammalian species, the question may be asked as to whether or not this is true throughout the mammalian scale. It has been well established that male and female rats which have been reared in physical isolation from others of their kind are completely capable of biologically effective mating when they are first put together at the time of female estrus (Beach, 1958). (See Fig. 5.4.)

At one time it appeared that at least some rodents cannot mate normally in adulthood if they are physically isolated from others of their kind during infancy. Valenstein and Young (1955) reported that male guinea pigs separated from their mothers and littermates a few days after birth have difficulty achieving intromission when placed with estrous females during adulthood. Initially this sexual inefficiency was interpreted as a result of prevention of practice in mounting other guinea pigs during prepuberal life. More recently this interpretation has been called into question by Gerall's finding (1962) that males reared in the same cage with other guinea pigs but separated from them by coarse-wire netting are able to mate normally in adulthood.

This factor is apparently also present in at least some domesticated Carnivora. The writer has conducted mating tests on male dogs which were reared in individual cages and allowed a minimum of physical contact with other dogs during puppyhood. Some of these animals never mated successfully because of their apparent inability to achieve intromission. They exhibited pronounced excitement and mounted the receptive bitch repeatedly, but the orientation was often inappropriate and the male might have persisted in mounting the side or head of the female instead of using the normal rear position.

In view of evidence cited earlier to the effect that male dogs and cats cease displaying sexual behavior following extensive injury to the forebrain, and at the same time comparable neurological lesions do not have this effect in the female, it is of considerable significance to note that females which have no sexual experience perform in perfectly normal fashion the first time they come into estrus and are placed with a sexually experienced male. In other words, individual practice is unnecessary as far as the female is concerned.

The most clear-cut evidence indicating the importance of sexual practice for the performance of the motor pattern of coition is derived from studies of monkeys and anthropoid apes.

Harlow (1962) has reared male and female rhesus monkeys without any contact with conspecific partners. When such animals reach adulthood and are tested as heterosexual pairs during the female's receptive period, they are unable to carry the coital pattern to a successful conclusion. The prime obstacle appears to be a failure of both partners to simultaneously assume the position necessary for the male's insertion. Although many copulatory attempts occur, intromission is not achieved.

Very much the same sort of difficulty is experienced by male and female chimpanzees which have been deprived of heterosexual contact during prepuberal life. Nissen (1954) studied a number of such cases and never saw successful copulation achieved by a naïve pair of apes.

When inexperienced males were placed with experienced females, intercourse was usually successful, but this was due entirely to the fact that the female made compensatory adjustments which overcame the male's inexpertness. A comparable situation obtained when naïve females were bred to experienced males. Even after they had successfully mated with experienced partners, males and females which had been reared apart from the opposite sex were only occasionally able to mate with one another.

The indicated conclusion is that, under normal circumstances, sex play occurring before puberty, under conditions of low drive, prepares the monkey and the chimpanzee for successful mating in adult life. The difficulty encountered by inexperienced adults in learning the proper bodily adjustments is probably related to the fact that at this age sexual drive is high and the resultant general excitement interferes with learning.

If one wishes to extrapolate and hazard a guess concerning the role of learning in human coitus it would have to be that "only practice makes perfect." Vicarious experience can undoubtedly provide a certain amount of guidance, but it appears reasonable to conclude that successful heterosexual intercourse between humans is heavily dependent upon individual experience and learning.

## Conclusions

Comparing rodents, carnivores, monkeys, apes, and humans we seem to detect three trends in the control of sexual activities. The first is a progressive decrease in the extent to which sexual arousal and performance are dependent upon gonadal hormones. The second is an increase in the degree to which the cerebral cortex influences and controls sexual expressions. The third is an increase in the importance of learning and experience for the successful execution of the sexual pattern of heterosexual union.

### REFERENCES

BARD, P. 1936. Oestrual behavior in surviving decorticate cats. *Amer. J. Physiol.*, **116:** 4–5.

BEACH, F. A. 1940. Effects of cortical lesions upon the copulatory behavior of male cats. *J. Comp. Psychol.*, **29:** 193–244.

———. 1942. Sexual behavior of prepuberal male and female rats treated with gonadal hormones. *J. Comp. Psychol.*, **34:** 285–92.

———. 1944. Effects of injury to the cerebral cortex upon sexually-receptive behavior in the female rat. *Psychosom. Med.*, **6:** 40–55.

———. 1945. Hormonal induction of mating responses in a rat with congenital absence of gonadal tissue. *Anat. Rec.*, **92:** 289–292.

———. 1947. A review of physiological and psychological studies of sexual behavior in mammals. *Physiol. Rev.*, **27:** 240–305.

———. 1958. Normal sexual behavior in male rats isolated at fourteen days of age. *J. Comp. Physiol. Psychol.*, **51:** 37–38.

Beach, F. A., *et al.* 1956. Experimental inhibition and restoration of mating behavior in male rats. *J. Genet. Psychol.*, **89:** 165–81.

Beach, F. A., and Fowler, H. 1959. Effects of "situational anxiety" on sexual behavior in male rats. *J. Comp. Physiol. Psychol.*, **52:** 245–48.

Beach, F. A., and Holz, A. M. 1946. Mating behavior in male rats castrated at various ages and injected with androgen. *J. Exp. Zool.*, **101:** 91–142.

Beach, F. A., Zitrin, A., and Jaynes, J. 1955. Neural mediation of mating in male cats, II: Contribution of the frontal cortex. *J. Exp. Zool.*, **139:** 381–402.

———. 1956. Neural mediation of mating in male cats, I: Effects of unilateral and bilateral removal of the neocortex. *J. Comp. Physiol. Psychol.*, **49:** 312–27.

Bingham, H. C. 1928. Sex development in apes. *Comp. Psychol. Monog.*, **5:** 1–180.

Brookhart, J. M., Dey, F. L., and Ranson, S. W. 1940. Failure of ovarian hormones to cause mating reactions in spayed guinea pigs with hypothalamic lesions. *Proc. Soc. Exp. Biol. Med.*, **44:** 61–68.

Brooks, C. McC. 1937. The role of the cerebral cortex and of various sense organs in the excitation and execution of mating activity in the rabbit. *Amer. J. Physiol.*, **120:** 544–53.

Dorfman, R. J. 1948. Biochemistry of androgens. *In:* G. Pincus and K. V. Thimann (eds.), *The Hormones*, pp. 467–548.

Filler, W., and Drezner, M. 1944. Results of surgical castration in women over forty. *Amer. J. Obstet. Gynec.*, **47:** 122–35.

Ford, C. S., and Beach, F. A. 1951. *Patterns of Sexual Behavior.* New York: Harper & Bros.

Gerall, A. A. 1962. Personal communication.

Goltz, F. 1892. Der Hund ohne Grosshirn. *Pflueger. Arch. Ges. Physiol.*, **51:** 570–85.

Green, J. D., Clemente, C. D., and DeGroot, J. 1957. Rhinencephalic lesions and behavior in cats. An analysis of the Klüver-Bucy syndrome with particular reference to normal and abnormal sexual behavior. *J. Comp. Neurol.*, **108:** 505–45.

Hagamen, W. D., and Zitmann, F. 1959. "Hypersexual" activity in normal male cats. *Anat. Rec.*, **133:** 388.

HARLOW, H. F. 1962. The heterosexual affectional system in monkeys. *Amer. Psychol.*, **17:** 1–9.

HELLER, O. C., NELSON, W. O., and ROTH, A. A. 1943. Functional prepuberal castration in males. *J. Endocr.*, **3:** 573–88.

HOOKER, C. W. 1944. The postnatal history and function of the interstitial cells of the testis of the bull. *Amer. J. Anat.*, **74:** 1–18.

KINSEY, A., POMEROY, W. B., and MARTIN, C. 1948. *Sexual Behavior in the Human Male.* Philadelphia: W. B. Saunder Co.

KINSEY, A. C., *et al.* 1953. *Sexual Behavior in the Human Female.* Philadelphia: W. B. Saunders Co.

KLÜVER, H., and BUCY, P. C. 1939. Preliminary analysis of functions of the temporal lobes in monkeys. *A.M.A. Arch. Neurol. Psychiat.*, **42:** 979.

KUEHN, R. E., and BEACH, F. A. 1963. Quantitative measurements of sexual receptivity in female rats. *Behaviour,* **21:** 282–99.

MACLEAN, P. D., and PLOOG, D. W. 1962. Cerebral representation of penile erection. *J. Neurophysiol.*, **25:** 29–55.

MICHAEL, R. P. 1961. Observations on the sexual behavior of the domestic cat (*Felis catus* L.) under laboratory conditions. *Behaviour,* **18:** 1–24.

MILLER, N. E. 1957. Experiments on motivation: Studies combining psychological, physiological, and pharmacological techniques. *Science,* **126:** 1271–78.

NISSEN, H. 1954. "Development of sexual behavior in chimpanzees." Unpublished record of conference on genetic, psychological and hormonal factors in the establishment and maintenance of patterns of sexual behavior in mammals, Amherst, Mass.

OLDS, J. 1958. Self-stimulation of the brain. *Science,* **127:** 315–25.

ROGERS, C. M. 1954. "Hypothalamic mediation of sex behavior in the male rat." Unpublished Ph.D. dissertation, Yale University Library.

ROSENBLATT, J. S., and ARONSON, L. R. 1958. The decline of sexual behavior in male cats after castration with special reference to the role of prior sexual experience. *Behaviour,* **12:** 285–338.

SCHREINER, L., and KLING, A. 1953. Behavioral changes following rhinencephalic injury in the cat. *J. Neurophysiol.*, **16:** 643–59.

SMITH, P. E., and ENGLE, E. T. 1927. Experimental evidence regarding the role of the anterior pituitary in the development and regulation of the genital system. *Amer. J. Anat.*, **40:** 159–67.

STEINACH, E., and KUN, H. 1928. Die entwicklungmechanische Bedeutung der Hypophysis als Aktivator der Keimdrüseninkretion. *Med. Klin.*, **24:** 524–37.

TERZIAN, H., and ORE, G. D. 1955. Syndrome of Klüver & Bucy reproduced in man by bilateral removal of the temporal lobes. *Neurology,* **5:** 373–80.

TINBERGEN, N. 1951. *The Study of Instinct.* Oxford: Clarendon Press.

VALENSTEIN, E. S., and YOUNG, W. C. 1955. An experiential factor influencing the effectiveness of testosterone propionate in eliciting sexual behavior in male guinea pigs. *Endocrinology,* **56:** 173–77.

YOUNG, W. C., and ORBISON, W. D. 1944. Changes in selected features of behavior in pairs of oppositely sexed chimpanzees during the sexual cycle and after ovariectomy. *J. Comp. Psychol.,* **37:** 107–22.

ZITRIN, A., JAYNES, J., and BEACH, F. A. 1956. Neural mediation of mating in male cats, III: Contributions of occipital, parietal and temporal cortex. *J. Comp. Neurol.,* **105:** 111–25.

DANIEL S. LEHRMAN

# 6

# Control of Behavior Cycles in Reproduction

In Chapter 4, we have seen that many vertebrates show long-term, or annual, cycles of changes in reproductive behavior. Within these cycles, it is often possible to see recurring shorter cycles of behavioral change, for example, in seasonally polyestrous mammals or in birds which produce several broods during a single breeding season. These behavioral changes are correlated with changes in the reproductive organs, the growth and development of which are controlled by the hormones secreted by the endocrine glands. The question therefore arises of whether, and how, cyclic changes in behavior associated with breeding are influenced by the endocrine glands. Many scientists have paid attention to this general problem, and we propose to devote this chapter to a discussion of the kinds of questions they have asked and to some of the answers which they have discovered. Additional information on mammalian mating patterns may be found in Chapter 5.

## Hormonal Induction of Reproductive Behavior

The first question to be dealt with is: Are the patterns of reproductive behavior in fact induced by hormones secreted by the animals' endocrine organs?

### SEXUAL BEHAVIOR

There is abundant evidence, for a great variety of higher animals, that sexual behavior depends upon the hormones secreted by the testis and the ovary. It will be recalled from Chapter 5 that removal of the testes

Daniel S. Lehrman is professor of psychology and director of the Institute of Animal Behavior, Rutgers University.

from a male rat or cat before puberty prevents the appearance of sexual behavior at the normal time of puberty and that the injection of male hormone into immature mammals may induce male sexual behavior at ages very much earlier than those at which it ever appears under normal circumstances; similar effects are found in females upon removal of the ovaries or injection of female hormones. Effects like these are also found in lower vertebrates. For example, young domestic chicks injected with male hormone may make the typical movements of a crowing cock, including stretching the neck, flapping the wings, etc., even when their vocal apparatus is so undeveloped that the accompanying vocalization is a high-pitched squeak! Similarly, injection of estrogenic hormones into immature female domestic chickens induces female mating behavior.

It is thus apparent that hormones produced by the testes and the ovaries play a significant role in inducing normal male or female sexual behavior. It must not be thought, however, that the expression of sexual behavior is a simple function of the presence of sex hormones or that quantitative variation in this behavior depends simply upon the amount of hormone present. This is far from the case. For example, the males of different strains of guinea pigs may show different amounts and intensities of sexual behavior when tested under identical conditions. If such animals are castrated, their sexual behavior disappears. If they are subsequently injected with male sex hormones, the original strain differences in level of sexual behavior ("sex drive") reappear even though all the animals received the *same* amount of sex hormone! Somewhat similar results are found in rats. If castrated male rats are injected with very small amounts of male sex hormone daily, it may be observed that increasing the amount of sex hormone injected causes increases in the level of sexual behavior, but only up to a certain point. Beyond that point, increasing the daily dose of male sex hormone has no further effect upon the amount of sexual behavior.

Now, if a number of mature rats are tested for sexual behavior with female rats, individual differences in the amount of sexual behavior will be seen; some animals can be scored "high," some "low," for the intensity of their sexual motivation. If these animals are castrated, the sex behavior will, of course, eventually disappear. If all the animals are then injected with the same high level of male sex hormone daily, sexual behavior will reappear. Under these circumstances, the animals which were "high" and those which were "low" before castration will, after injection with sex hormone, return to approximately the same level of sexual motivation as that which they had before castration (Beach, 1948; Young, 1961).

Data like these indicate that individual differences in the amount of sexual behavior in different animals of the same species can probably not be accounted for primarily by individual differences in the amount of sex hormone being secreted by their own glands. It appears far more likely that all mature animals secrete more than enough sex hormone to induce normal sexual behavior and that individual differences in the amount of sex behavior reflect differences in the sensitivity of the animals' tissues to the sex hormone, rather than differences in the amount of sex hormone present. The nature of these differences in tissue sensitivity is still quite obscure. Some individual differences depend upon the individual experiences which the animals have had, and we shall discuss these later.

Not only individual differences but also species differences may be seen in the response of different animals to injection of the same kind of hormone. Injection of testosterone propionate into an immature or castrated rat, cat, guinea pig, dove, or chicken will, in each case, have the effect of inducing the appearance of the male sex behavior characteristic of the species. The ring dove will bow and coo at a female; the rooster will crow and attempt to tread a hen; the cat will grip the neck of a female cat; the rat will sniff at the genitals of a female rat. These behavior patterns look very different from each other, in spite of the fact that they were all induced by exactly the same treatment. This is also true of the natural situation. The male sex hormone secreted by the testes of different kinds of higher animals is chemically similar or identical in the different species, even though the organization of the sexual behavior pattern induced by it (like the anatomy of the sex organs influenced by it) varies very widely. This implies the existence of some degree of continuity in some of the basic physiological mechanisms controlling sexual behavior, co-existent with the inter-species discontinuities in the patterns of behavior which have been produced by evolution.

### PARENTAL BEHAVIOR

Parental behavior, as well as sexual behavior, appears to be influenced by endocrine secretions. If prolactin (a hormone normally secreted by the pituitary gland in certain phases of the reproductive cycle) is injected into domestic chickens which are laying eggs, they will stop laying and start to incubate the eggs in the nest. The same hormone injected into non-laying hens or into capons or roosters does not induce them to sit on eggs but may sometimes induce them to show parental behavior toward young chicks.

The same hormone, secreted by the pituitary glands of pigeons and doves while they are sitting on eggs, prepares them to feed their young after hatching by regurgitating to them a substance ("crop milk") produced in the crop. Injection of prolactin into ring doves who have had experience in rearing young induces them to feed young doves placed in the cage with them. In the case of ring doves, the hormone which appears to induce the beginning of incubation behavior is progesterone, secreted by the ovary at about the time that the eggs are laid (Lehrman, 1961).

Many mammals, including rats, mice, rabbits, etc., build nests, either during pregnancy or at the end of pregnancy. In some of these cases, it has been demonstrated that the special conditions of hormone secretion during pregnancy or around the time of parturition play a role in the induction of this nest-building behavior.

Mice of either sex regularly do a small amount of nest-building activity, regardless of pregnancy. In the domestic mouse pregnant females do very much more nest-building than non-pregnant females, males, or immature animals—the nest-building activity increasing abruptly several days after conception. Since this is also the time when the ovary begins to secrete the large amounts of progesterone which are characteristic of pregnant mammals, Koller (1952, 1956) investigated the effects of this hormone on nest-building behavior. Under the influence of injected progesterone, immature mice or females whose ovaries had been removed did just as much nest-building as do pregnant females.

The situation in the pregnant rabbit appears to be quite different from that in the mouse. To begin with, nest-building in the rabbit occurs just before or just after parturition, rather than during most of pregnancy, as in the mouse. Administration of progesterone does not induce nest-building in the rabbit, but treatments resulting in the *cessation* of progesterone secretion, such as removal of the uterus and cervix in pregnant animals, do induce nest-building. Administration of combinations of estrogen and progesterone in various doses and for varying times shows that nest-building in this animal is induced by a sudden reduction in the amount of progesterone in the blood in relation to the amount of estrogen. This reduction normally occurs at the time of parturition, when the expulsion of the placenta removes a major source of progesterone.

It is apparent from these remarks that, when dealing with parental behavior, as with sexual behavior, we must keep in mind that the pattern of behavior induced by various hormones may vary significantly in different kinds of animals, sometimes even in animals fairly closely related to each other. We may conclude from all these facts that the cycle of changing patterns of behavior associated with courtship, nest-building, care of the young, etc. are, at least to some extent, actually induced by the changing endocrine conditions during a breeding cycle.

## Mechanisms of Hormonal Induction of Behavior

The second question we must ask is: How do hormones act in an organism, so as to change its behavior?

When a hormone causes the appearance of, or any changes in, a complex behavior pattern, this naturally means that the administration of the hormone has caused some changes in the activity of the central nervous system. But there are many ways in which such changes could be brought about. The hormone might have a general effect upon the excitability of the nervous system; it might have specific effects upon various structures of the nervous system; or the hormone might have effects on structures outside the central nervous system which could cause changes in the flow of sensory information into the central nervous system, thus changing its activity. It requires careful experimentation to distinguish among these alternatives and to discover how, in any particular case, a hormone brings about a behavioral effect.

### PERIPHERAL HORMONE EFFECTS

When a ring dove with previous breeding experience is injected with prolactin, its crop wall thickens until the crop weighs several times as much as it does in its resting condition. The superficial layers of the crop lining slough off, so that the crop becomes filled with a cheesy mass of degenerating epithelial cells, which constitutes the "crop milk" which the birds regurgitate to their dependent young. If a ring dove in this condition is placed with a young ring dove, it will feed it by approaching it, gently pecking at its bill so the squab thrusts its head into the parent's throat, then regurgitating to it. If an experienced adult ring dove which has *not* been injected with prolactin is placed in exactly the same situation, it does not feed the young, nor does it show any parental behavior toward it, such as approaching it, attempting to brood it, etc. Obviously, the prolactin has induced this parental feeding behavior. How has it done so? There are two general possibilities: The prolactin may have acted upon some structure in the central nervous system so as to arouse there the motive to feed the young, or there may have been some central nervous change in perceptual or motor organization which would have this motivational effect; or, the effect of the hormone somewhere outside the central nervous system, such as its effect in engorging the crop, might cause a change in the inflow of nerve impulses *to* the central nervous system, thus inducing those activities in the brain which result in this form of parental behavior.

If the wall of the crop is anesthetized with a local anesthetic, then the fact that it is engorged cannot have any effect upon sensory inflow to the brain. If the effect of prolactin in inducing parental feeding behavior were *via* its effect on the crop, then locally anesthetizing the crop should prevent prolactin-injected birds from showing this parental approach to young squabs. If, on the other hand, prolactin aroused this parental behavior by a direct chemical effect on some "center" in the brain where this kind of behavior is organized, then anesthetizing the crop should not prevent prolactin-injected birds from approaching the young, although it might, of course, prevent actual regurgitation by paralyzing some of the muscles involved.

Local anesthesia of the crop does, in fact, inhibit parental feeding behavior, indicating that the behavioral effects of prolactin in this situation are, at least in part, mediated by its effect upon the crop. Another way in which prolactin contributes to the occurrence of parental feeding behavior is by suppressing the secretion by the pituitary gland of gonad-stimulating hormones, so that the testis and ovary do not secrete sex hormones. This results in the suppression of sexual behavior, which would interfere with the expression of parental behavior. Since the pituitary gland is, at least in part, controlled by the brain, *this* effect of prolactin on pituitary secretion might well be through an effect upon the nervous system but *not* upon a part of the nervous system specific for parental behavior.

From this example, it can be seen that a hormone may sometimes induce a change in behavior by causing several different kinds of effects, all of them necessary for the behavior to appear, and none of them necessarily constituting a direct hormonal effect upon a central nervous mechanism for that behavior.

There are other examples of hormonal changes induced in peripheral structures which *may* be relevant to the induction of behavioral changes, but specific evidence is lacking. For example, many species of birds, when they incubate eggs, do so by pressing against the eggs an "incubation patch" which is a defeathered, thickened, vascularized area of skin on the underside of the bird's body. The loss of feathers, thickening of the epithelium, and growth of blood vessels usually occur just at about the time when the birds begin to sit on the eggs, and these changes are brought about by a co-operative effect of sex hormones and prolactin. It can be demonstrated that the ventral skin of canaries changes in tactual sensitivity during this time, and it is possible that sensory changes, induced by hormones, in the incubation patch play a role in the motivation of the bird to sit on its eggs. Even if this is true, it is not known how important a role it has or what other factors might be involved.

Similarly, the tendency of female mammals to show nursing behavior toward their young seems to be closely correlated with the presence of milk in their mammary glands. The accumulation of milk causes increases in tension in the mammary glands, and the nursing experience itself induces increases of pressure in the milk-secreting tubules. It has been shown, for example, that female rabbits, who only visit their young once a day to nurse them, do so at a time when mammary tension is high. The waxing and waning of nursing behavior during the period after birth of the young is normally closely correlated with the production of milk. The production of milk depends upon stimulation of the mammary glands by ovarian hormones and by prolactin and other pituitary hormones, and it seems likely that, to the extent that these hormones induce maternal nursing behavior, they may do so partly by their effect on tensions in the mammary glands. There are, however, situations under which some mammals will attempt to show nursing behavior when they are not lactating, and it is not yet known just what role tensions in the mammary glands play in the motivation of nursing behavior.

### CENTRAL BEHAVIORAL EFFECTS OF HORMONES

In recent years, a number of lines of evidence have strikingly converged toward the conclusion that the sex hormones can influence the appearance of sexual behavior patterns by means of direct effects of the hormones upon the central nervous system.

A female cat whose ovaries have been removed may be made to show complete female sexual behavior by the injection of estrogenic hormone. This hormone, however, not only causes the appearance of receptiveness to the sexual approaches of a male cat but also causes the growth and development of the sex organs (vagina, uterus, oviduct, etc.). The question therefore arises whether the female shows this sexually receptive behavior because of the direct effect of the hormone upon her brain or because of changes in sensation in the sex organs induced by the hormones. If we could inject the hormone in such amounts that it caused the sexual behavior without causing any changes in the sex organs, this would indicate that the hormone might act on the central nervous system or, at any rate, somewhere else than on the reproductive organs. Unfortunately, the amount of estrogenic hormone which must be injected in order to induce sexually receptive behavior is *greater* than that required for the growth of the sex organs, so this experiment cannot be done merely by manipulating the amount of sex hormone injected into the blood system. It is, however, possible to implant minute amounts of estrogenic hormone directly into the brain. If a fine needle is dipped into a fatty-acid form of estrogenic hormone, it will become coated with a very thin layer, which may be limited to the point itself, of female sex hormone in a form which dissolves very slowly in body fluids. If this needle is implanted directly into the hypothalamus, so that the hormone comes into contact only with one small part of the hypothalamus without affecting any neighboring nerve tissue, full estrous behavior may be seen without any growth whatever of the reproductive organs. Needles without the hormone coating have no effect; implantation of the needle in other, even nearby, regions of the brain also has no effect. This experiment strongly suggests that sex hormones may influence sexual behavior patterns by local effects, directly in the brain, on structures concerned with the organization of the behavior. There is even evidence, obtained by tracing radioactive molecules of estrogen, that specific neurons of the hypothalamus may have an affinity for these molecules. There are still, of course, a number of problems. We do not know whether other kinds of steroid hormones would have similar effects; we do not know what the effects of such local implantations of female sex hormones would be in the corresponding locus in the hypothalamus of a male cat; we do not know why, for example, local implantation of male sex hormones in the hypothalamus of a male rat should cause it to display behavior like retrieving of the young, which is not induced by injecting the same hormone into the blood system; we do not know to what extent effects of the hormones on levels of excitability in the nervous system at places other than the specific loci concerned in these experiments may contribute to the arousal of sexual behavior. But it is clear that direct effects of gonadal hormones on the nervous system play an important role in the arousal of sexual behavior.

Let us consider how the organization of sexual behavior is influenced during the early development of the nervous system. Male and female guinea pigs characteristically have different patterns of sexual behavior, and it is possible to develop quantitative measures for the degree of maleness or femaleness to be seen in the behavior of any particular animal in a standard test situation (Young, 1961, 1962). If male sex hormone is injected into a pregnant female guinea pig and she is then allowed to give birth, it is possible, by studying the sexual behavior of the offspring guinea pig after it reaches maturity, to determine what effect is exerted upon adult sexual behavior by the sex hormone administered only during fetal life. The results are quite striking. All sorts of variations in sexual behavior, from the full female pattern all the way to an almost completely male-like pattern of sexual behavior, occur among genetically female offspring exposed to male hormone before birth. In addition to this behavioral "intersexuality," these animals also show ambiguity in their sexual anatomy. The reproductive organs are intermediate between female- and male-like, just as is the sexual behavior. However, the degree to which the sexual behavior is distorted in the male direction in any particular animal does not necessarily correspond with the degree to which the sexual anatomy is distorted in the same animal. This indicates that the distortion of behavior in the male direction is not simply a consequence of the distortion of the anatomy of the sex organs in the male direction and suggests that the behavioral effect of the hormone administration might have been on the development of the central nervous system. The following bit of evidence points in the same direction: Although the ovaries appeared to be perfectly normal up to the time of puberty, many of these treated animals had irregular and abnormal estrous cycles. Since the hypothalamus controls the cyclic activity of the pituitary gland (which, of course, controls the ovarian cycle), and since the hypothalamus, independently of this effect upon the pituitary gland, also plays an important role in the organization of sexual behavior patterns, it is fairly clear that the administration of male hormone during fetal life in female embryos results in a distortion of the genetically expected direction of development, so that the hypothalamus develops in a male direction. This suggests not only that the behavioral effects of sex hormones are probably on the central nervous system but also that the hypothalami of normal male and female animals of the same species are different from each other in ways which are relevant both to the manner in which the hypothalamus controls the pituitary gland differently in the two sexes, and to the way in which sexual behavior is different in the two sexes of the same species.

If the fetuses are genetically male, the injection of male hormone has no effect upon their development. It has not yet been found possible to test the effects of female sex hormone injected during fetal life on the sexual development of genetically male guinea pigs. This is because female sex hormone injected into the pregnant guinea pigs does not get into the fetus, being barred from it by the same protective mechanisms which prevent the mother's own sex hormones from unduly affecting the fetus.

It is clear that many aspects of the development and the expression of sexual behavior depend upon direct effects of sex hormones on the nervous system.

## Environmental Stimulation of Hormone Secretion

The third question which we wish to ask is: What makes the glands secrete their hormones just at the right time in the reproductive cycle for them to induce behavior appropriate to that stage of the cycle?

As explained in Chapter 2, the pituitary gland is not only located at the base of the brain but is actually connected to the brain in such a way that the hypothalamus can exert specific and detailed control over many of its secretory functions. Since secretion of sex hormones by the testis or ovary depends upon stimulation by gonad-stimulating hormones from the pituitary gland, the brain-pituitary link provides a mechanism for control of the secretion of sex hormones by the hypothalamus. The fact that the brain can control the pituitary gland suggests that stimuli arising from the animal's environment or within its body can stimulate the secretion of hormones by neural mechanisms analogous to, or identical with, those by which the stimuli elicit behavioral responses. There is a great deal of evidence that this is indeed the case and that environmental regulations play a substantial and important role in the organization of reproductive cycles.

Temperate-zone birds breed in the spring, and it has long been known that their spring migration, courtship singing, and reproductive activities are associated with growth of the gonads and secretion of sex hormones at that time of year. If birds are kept in the summer and fall, when the days are getting shorter, in an aviary supplied with artificial lighting and the lights are turned off and on at such times so that the days appear to be getting longer instead of shorter, the testes and ovaries will grow as they do in the spring, instead of regressing as they do in the late summer and fall. Many of the physiological signs of spring breeding conditions can be thus induced in the fall: growth of the gonads, singing, migratory restlessness, pre-migratory deposition of fat, etc.

Experiments like these indicate that the pituitary gland is stimulated by the increasing length of the day to produce gonad-stimulating hormones. Note that this is, in effect, a "reflex" secretion of hormone in response to an external stimulus.

The length of the day is not the only external stimulus which induces the endocrine changes associated with the breeding season. In some parts of the world, such as in the tropics, the length of the day does not change appreciably at different times of year; in other places, such as desert areas with irregular rainfall, breeding is sometimes associated with the sporadic occurrence of rain, rather than with any particular time of year. The African weaver-finch, *Quelea*, for example, breeds after rain, and it has been demonstrated experimentally that the testes of the males are induced to develop by the presence of green grass, which they use for nest-building. Neither rain itself nor the presence of insect food (another consequence of rain) induces gonadal growth in the absence of green grass; green grass, introduced into the birds' cages, induces gonad growth in the absence of either rain or insect food. Females of this species undergo changes of bill color, indicating pituitary stimulation, simply as a result of seeing the male build a nest!

The oviducts of female ring doves, tested in the laboratory, increase in weight when the female is placed with a sexually active male (Fig. 6.1). This effect occurs even if the male and female are separated by a glass plate, indicating that the growth of the female's oviduct and, therefore, the secretion of oviduct-stimulating hormones by her ovary, are stimulated by visual (and auditory) stimuli from the male. Field ornithologists have long known that, in many species of birds, the female does not permit copulation or start to build the nest until after some time (up to three or four weeks) of persistent courtship by the male. It appears that the behavior of the courting male actually induces endocrine secretion in the female which causes her to display sexual behavior and to build the nest.

The nest, in turn, provides stimuli which induce further changes in endocrine secretion. Female ring doves placed with a male in a cage

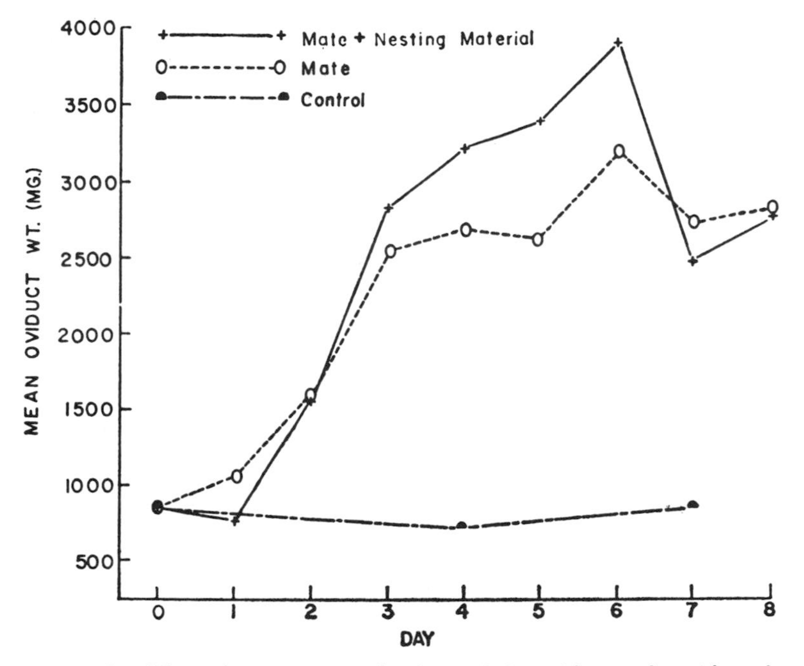


FIG. 6.1. Effect of varying periods of association with a male, with and without the presence of nesting material, upon the oviduct weight of female ring doves. The controls were placed alone in breeding cages. (From Lehrman, Brody, and Wortis, 1961.)

without a nest-bowl or nesting material do not produce eggs so quickly or become ready to sit on eggs so early as do females placed in a cage with a male *and* nest-bowl and nesting material (Fig. 6.2). This means that stimuli coming from the nest-bowl and nesting material or from the consequent participation in nest-building behavior induce hormone secretion additional to that induced by the male alone. Anatomical study of the changes in the oviduct during this time indicates that stimuli coming from the nest-bowl and nesting material facilitate the secretion of progesterone and the laying of the eggs. Endocrine changes appear to occur in the male as well, as a consequence of his association with the female, since his readiness to build a nest and his later readiness to sit on the eggs (both sexes sit in this species) develop gradually during his association with the female. These changes have not yet been analyzed.

In wild birds, too, it almost never happens that an egg is laid before

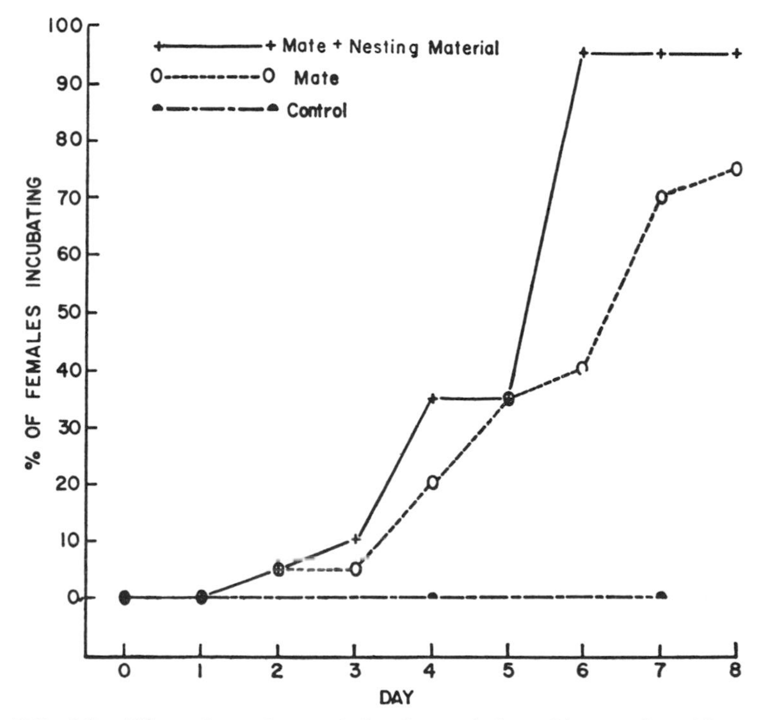


FIG. 6.2. Effect of varying periods of association with a male, with or without the presence of nesting material, upon the development of the female ring dove's readiness to sit upon eggs. The birds were tested for their response to a nest and eggs introduced into the cage after various periods in the cage alone (control), with a male (mate), or with a male plus nesting material. (From Lehrman, Brody, and Wortis, 1961.)

a nest is built. If a nest-building female suffers the loss of a nest while she is building, through the activities of a storm or of an experimenter, she will not lay her eggs until she has finished building the new nest which she must then start. This means that the eggs will not be laid until two or three days after the time when they would ordinarily have been laid. The implication is clear: stimuli coming from the nest facilitate the production of the eggs, which means that they stimulate the secretion of ovary-stimulating pituitary hormones.

The activity of the pituitary gland at this time is also influenced by stimuli arising within the bird's body, although outside of its central nervous system. The ovary of a domestic chicken releases a new egg into the oviduct within an hour after the laying of the previous egg. If the laying of the previous egg is delayed by putting the hen in an unsuitable situation, the ovary will still not release the next egg until the preceding one has actually been laid. How is this accomplished? Huston and Nalbandov (1953) sewed a loop of thread into the wall of the upper portion of the oviduct in a number of hens. Most of these hens laid few or no eggs as long as the loop of thread was left in place. As soon as the loop ware removed, the hens began to lay eggs at the normal rate. The mechanical tension exerted on the oviduct wall by this loop of thread duplicates that exerted by an egg in the oviduct; this experiment means that the pressure exerted by an unlaid egg in the oviduct is a stimulus which prevents the pituitary gland from releasing its ovulation-inducing hormone. During the time when the loop of thread was in place, the oviducts of the hens did not decrease in weight. This means that the mechanical stimulus provided by the thread, which prevented the secretion of one of the ovary-stimulating pituitary hormones, did *not* prevent the secretion of other pituitary hormones which make the ovary secrete the hormone which maintains the weight of the oviduct. Thus, specific stimuli can affect the secretion of particular pituitary hormones, without necessarily affecting the secretion of other hormones from the same gland.

Once the ring doves begin regular incubation, under the influence of the progesterone secreted as a result of their participation in nest-building, their pituitary glands begin to secrete prolactin, as shown by the fact that their crops gradually become heavier during the incubation period. If the birds are removed from the breeding cage and thus separated from their nest at any time during the incubation period, the crop quickly returns to its resting weight, indicating the cessation of prolactin secretion by the pituitary gland. If a male pigeon is removed from the breeding cage but, instead of being completely separated from his mate, is allowed to stay in a neighboring cage from which he can see her sitting on the eggs, then his crop will develop just as if he were sitting on the eggs himself! This indicates not only that stimuli associated with the act of incubation induce the secretion of prolactin by the pituitary gland but that even *visual* stimuli, not involving any contact with the eggs, are able to induce and maintain this hormone secretion.

The reproductive cycle of birds thus appears to be continuously subject to environmental influence. The changing length of the day or changing ecological conditions provide stimulation which induces a reproductively active condition of the gonads, at an appropriate time of year. Within this time of year or breeding season, successive changes in the endocrine condition, which induce the successive changes in behavior, are themselves induced by stimuli coming from the environment, including those arising from the behavior of the bird and of its mate, such as song, the nest, eggs, etc.

In mammals, too, external stimuli play an important role in the regulation of endocrine secretion. Laboratory rats and mice come into estrus regularly every four or five days, and it is well known that this cyclic variation in ovarian condition depends upon a set of reciprocal exciting and inhibiting relationships between the pituitary gland and the ovary. However, this reciprocal interaction between the pituitary and the ovary, in which each secretes hormones having an effect upon the other so that they repeatedly alternate in function, is not entirely independent of external stimulation. Female rats all tend to come into estrus at approximately the same time of day (or rather, of night). This is because the alternation of light and dark is a stimulus to which the pituitary gland tends to respond by secreting an ovulation-inducing hormone, which, in turn, stimulates the ovary. As a result of the cycle of interaction between the pituitary gland and the ovary, it is only at intervals of several days that the pituitary-ovary system is in such a condition that the external stimulus associated with the alternation between day and night *can* induce this effect, but the importance of environment should nevertheless not be overlooked.

The female rabbit is more or less constantly in estrus, but she does not ovulate until after copulation with a male. Approximately the same situation obtains in the female domestic cat during her periods of heat; she ovulates only after copulation with a male. In both these animals, injection of an ovulation-inducing pituitary hormone is capable of causing the release of an egg by the ovary, even though no copulation has occurred. The cat can be induced to ovulate by mechanical stimulation of the vagina without the presence of a male cat. Further, direct electrical stimulation of the hypothalamus can also induce ovulation. Clearly, ovulation in these animals results from the reflex secretion of an ovary-stimulating hormone by the pituitary gland in response to stimulation associated with participation in copulation.

Female mice normally become pregnant after a single copulation with the male. If a female mouse is placed with a strange male mouse (i.e., a male mouse other than the one with which she has mated) within a day or two after copulation, the presence of the strange male may prevent pregnancy from becoming established. By using male mice very closely related with the one with which she mated (e.g., littermates in an inbred strain) or less closely related (e.g., mice of a different strain), it has been demonstrated that the more different the strange male mouse is from the mouse with which the female mated, the more likely it is that his presence will interfere with the establishment of pregnancy, regardless of whether the mouse with which she mated belonged to the same strain as, or to a different strain from, the female! (See Fig. 2.5.)

This block to the establishment of pregnancy depends entirely upon olfactory stimulation from the strange male, as is shown by the fact that the male need not be visible to the female and by the fact that a box in which the strange male lived until recently may be just as effective in blocking pregnancy as is the strange male itself. The effect of the olfactory stimulus of the strange male appears to be to block the secretion of prolactin (which is involved in the establishment of pregnancy) by the female's pituitary gland (Parkes and Bruce, 1962).

The establishment and maintenance of lactation by female mammals after the birth of their young is an excellent example of the way in which internal and external environments co-operate to control the endocrine system. The mammary glands of a pregnant mammal develop under the influence of estrogen and progesterone produced by the ovary and the placenta, so that they are capable of secreting milk by the time of parturition. Milk is not actually secreted until just after parturition, apparently as a consequence of the hormonal changes resulting from the expulsion of the placenta. After parturition, the pituitary gland secretes increased amounts of prolactin, and the mammary gland consequently starts to produce milk. If the young are removed from the mother at birth, this milk secretion dries up after a day or two in most species of mammal, and estrus cycles are shortly resumed. If the young remain with the mother, and she nurses them, milk secretion is maintained until the young are old enough to find food independently.

Now, if the young mice or rats are removed from the mothers some time before weaning age and replaced with foster-young, and if this is done repeatedly, so that the mother continues to have young of an age to engage in active suckling behavior, the period during which the mother produces milk may be prolonged until it is four or five times as long as the normal lactation period. This implies that stimuli coming from the suckling young actually induce secretion by the pituitary glands of the hormones which make the mammary glands secrete milk. If young rats are removed from their mothers in the middle of a lactation period, lactation rapidly declines, and the mammary glands return to an inactive condition. This decline can be delayed in *all* of the mammary glands by painting *some* of the nipples with turpentine. The turpentine, by irritating the tissues on which it is painted, causes continuous arousal of the receptors normally stimulated by the suckling young. If some of the nipples of a rat are denervated, stimulation of those nipples does not result in the maintenance of milk secretion. However, stimulation of the remaining (non-denervated) nipples induces milk secretion even in those mammary glands which were denervated. This indicates that the stimulus for the maintenance of milk secretion is conveyed from the nipple to the central nervous system by nerve connections but that the influences which, in turn, reach the mammary gland in order to maintain secretion do not require a nerve connection. This means that they are probably carried to all the mammary glands by the blood.

These and numerous other experiments indicate that the suckling stimulus elicits a reflex secretion of prolactin by the pituitary gland, which maintains the mammary gland in a secretory condition. Furthermore, the active expulsion of milk by the mammary gland, which occurs at each suckling episode, results from the stimulation of the mammary tissues by a second hormone reflexly released from the pituitary gland in response to suckling stimulation.

All these data indicate that the sequences of stages of endocrine secretion, inducing changes in behavior, in birds and in mammals (and in other animals as well), represent a remarkably intimate and effective interaction between the inner and outer environments. The hormone secretions induce behavioral states, which are states of responsiveness to particular aspects of the environment. The environment, in turn, provides stimulation, to which the animals are now sensitive, which causes further changes in endocrine secretion, which are responsible for still further changes in the animal's behavior. If a relevant part of the environment is a mate which is itself changing in physiological condition or young produced by the behavior of the animal, it can readily be seen that the synchronization of the physiological conditions and behavioral states between two members of a pair of birds or the synchronization of the maternal behavior and milk-producing activity of a mother mammal to the characteristics of her developing young reflect a constant adjustment of the animal's endocrine system to the changing conditions of its external environment.

## The Role of Experience in the Development of Hormone-Induced Behavior Patterns

The fourth question which we wish to ask is: What role does the animal's individual experience play in the development of behavior patterns which depend upon hormones?

At first sight, it might appear that the facts (*a*) that a behavior pattern could be induced by stimulation of a particular part of the brain, (*b*) that it appears more or less normally the first time the animal breeds, and (*c*) that it occurs in the same way in all members of the species imply that individual experience could not play any role in its development or organization. But, as will be seen, such a conclusion would be unjustified. No amount of knowledge of the *structure* of a behavior or of the mechanisms giving rise to it in the adult animal can justify any confidence in statements about the way in which it developed. It is only by studying the development itself that we can gain insight into the processes through which the behavior has become established.

There are many examples of hormone-induced behavior patterns in which it can be demonstrated that animals with different kinds and amounts of experience react very differently to the injection of a hormone. For example, we mentioned earlier that progesterone induces ring doves with previous breeding experience to sit on eggs. A comparison of the behavior of ring doves with and without previous breeding experience is very illuminating. If experienced, progesterone-injected birds are introduced into a cage containing a nest and two eggs, half of them will be sitting on the eggs within twenty-two minutes, and all of them within three hours. If exactly the same treatment is applied to birds of the same age but without previous breeding experience, it is twenty-four hours before half the birds are sitting on the eggs, and some of the birds never sit! This difference in behavior is apparent at the very beginning when the birds are introduced into the nest. The birds with previous breeding experience immediately go to the neighborhood of the nest; all of them are standing near the nest within one minute after being introduced into the cage. None of the inexperienced birds stand near the nest in less than three minutes, and many take more than two hours.

Similar data are obtained in comparing ring doves with and without previous breeding experience in their reaction to young doves after prolactin injection. Birds with previous breeding experience go to the young and feed them in less than an hour, inducing the young birds to feed by pecking them gently around the head. Inexperienced birds subjected to the same hormone treatment do not approach the young and do not feed them.

When male guinea pigs are reared under conditions of association with other animals of the same age, they develop higher scores on various measures of the intensity and efficiency of sexual activity than do similar guinea pigs reared in social isolation. This indicates that the kind of early experience which the animal has had contributes to the efficiency of its sexual behavior. Remember that this sexual behavior is induced by male hormone administration and eliminated by castration. It appears, therefore, in the case of sexual behavior in guinea pigs as in the case of parental behavior in the ring dove, that the kind of experience the animal has had influences the behavioral response to the hormones. This interaction between the effects of early experience and the effects of hormone administration can be shown directly. If one group of guinea pigs is raised in groups, or "socially," and another group raised in individual cages, the result will be, as just stated, that the animals reared socially will be more efficient in the performance of sexual behavior than will the animals reared in isolation. If all of these animals are then castrated, the sexual behavior score will go down to a minimal level for both groups. If, several months after castration, all the animals are treated with identical amounts of male sex hormones, sexual behavior will be reinstated in both groups, but the level of sexual behavior reached by each group will be that characteristic of its own group. That is, the animals reared in groups will still have a higher level of sexual behavior than the animals reared in isolation (Fig. 6.3).

This experiment indicates that the sexual behavior pattern induced by male hormones after maturity is, in part, dependent upon the experience which the animals had before they were sexually mature.

Some analogous results can be seen in experiments with male cats. If male cats are castrated before puberty, they never develop sexual behavior. If they are castrated after puberty, sexual behavior may persist for some time after castration. If cats are put in experimental cages before puberty and then kept until after they are sexually mature, they may be divided into two groups, one of which will be allowed to have sexual experience after puberty, the other not. Now, if cats of both these groups are then castrated and tested for sexual behavior by being placed with estrous female cats at intervals after castration, a striking difference becomes evident between those cats which did and those which did not have sexual experience before castration (Fig. 6.4). The sexually experienced cats show a much higher level of sexual behavior after castration than do the inexperienced ones. Although the

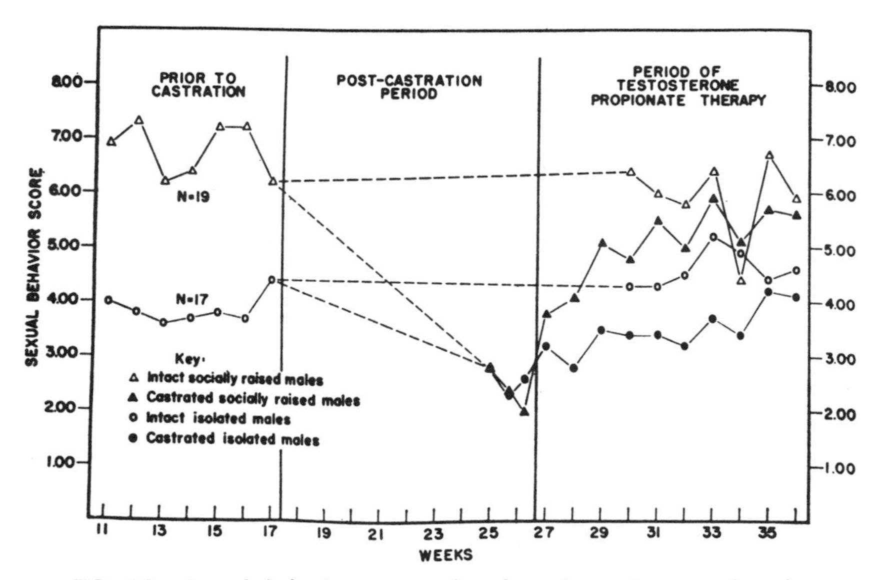


FIG. 6.3. Sexual behavior scores of male guinea pigs reared under different conditions, castrated, and then later injected with male hormone (see text). (From Valenstein and Young, 1955.)

sexual behavior of the experienced cats wanes somewhat after castration, it does not, even after several months, reach as low a level as that of the inexperienced cats. However, if these cats are kept for very long times (several years) after castration, the sexual behavior even of the experienced animals declines to a very low level. If male hormone is then administered to these animals for a period of several months and then withdrawn, subsequent tests show that the animals which originally were experienced are now able to perform sexually at a level comparable to that which they showed after their original castration. No such effect is seen in the originally inexperienced animals. This means that, although sexual behavior cannot develop in these animals unless in the presence of male sex hormone, once it has developed it represents a very persistent change in the animal's organization, the effects of which interact with those of sex hormones at later periods in the animal's life.

All of these experiments show that the effects of experience can interact with those of hormones in many different ways. Some relevant experience occurs before sexual maturity; some occurs after sexual maturity. Some relevant experience is experience of the actual behavior;

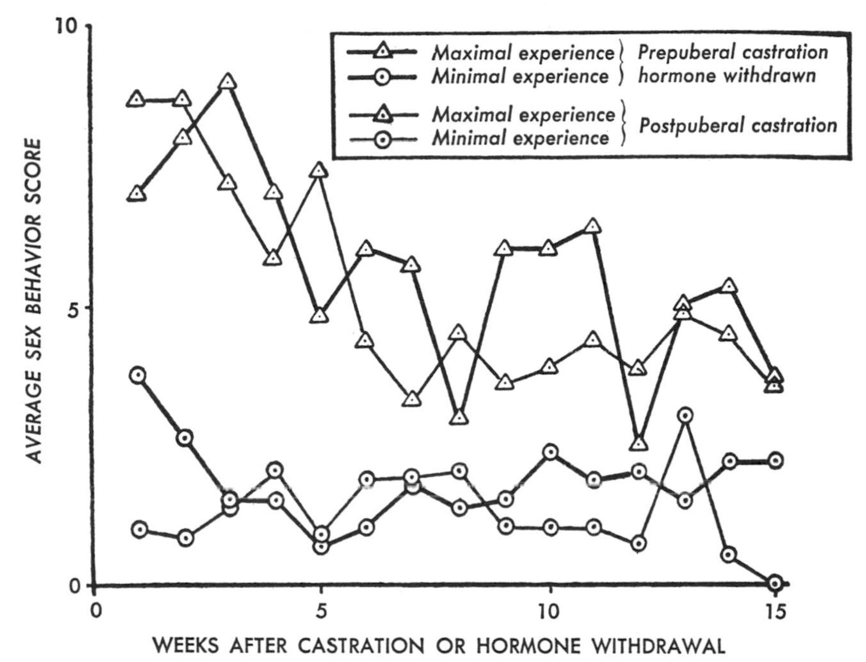


FIG. 6.4. Sexual behavior after castration, or after withdrawal of hormone administration, in male cats with and without previous sexual experience (see text). (From Rosenblatt and Aronson, 1958.)

other relevant experience may be experience, not of the actual behavior, but of a kind which facilitates the development of the behavior.

If ring doves are so strikingly less able to sit on eggs or to feed the young in response to injection of the appropriate hormones when they have not had previous breeding experience than when they have had this experience, how does it happen that doves breeding for the first time manage to do these things appropriately? The answer lies in the fact that we must not think of "experience" as being limited to the effects of having performed the same behavior before. When a dove builds a nest and thus becomes attached to the nesting site so that she spends most of her time there, she is acquiring "experience," through which she becomes oriented to the nest in such a way that (*a*) the nest is the place where she is most likely to lay the egg, and (*b*) she is bound to come into contact with the egg after it appears even without having any intention to sit on it. Similarly, when doves sit on the eggs, they become attached to the nest-site even more, and they are there when the eggs hatch, so that they come into contact with the newly hatched squabs. When a pregnant cat licks itself, it is learning something about the tastes and smells of its body which may help to orient it to the young. When it licks the young during and just after parturition, it is developing a connection with them through which it becomes easier to establish the nursing relationship. When it nurses the immobile young, it develops an attachment to them which makes it possible to maintain the nursing relationship even after the kittens are able to move about so that they are not in compulsory contact with the mother. All of these events have in common the fact that the animal's experience during early stages of the reproductive cycle may have an effect on its orientation to the world and upon its responses during later stages of the *same* cycle. This means that experience is playing a role in the succession from stage to stage even during the first breeding cycle. It is possible, therefore, that the first breeding cycle may be as efficient, or almost as efficient, as successive breeding cycles, while, at the same time, injection of hormones which project an animal directly into the middle of a breeding cycle may find experienced animals capable of orienting themselves toward the appropriate object of the behavior induced by the hormone and inexperienced animals incapable of doing so.

## Conclusion

In this chapter we have attempted to formulate some of the more important problems concerned in the analysis of hormone-behavior relationships in the regulation of breeding cycles and to suggest some of the directions in which answers are being sought to these problems. It should be apparent that the analysis of relations between hormones and behavior is not a simple matter of studying the effect of a simply-defined set of substances upon equally simply-defined resultant processes. The study of hormone-behavior relationships takes us into the most basic problems involved in understanding the organization of animals and their relationships with their environment: the structure of behavior, the nature of developmental processes, the organization of the nervous system, the functional links between the nervous system and the endocrine system, relationships between the inner environment and the outer environment, adaptive variations in patterns of life and physiological processes in animals living in different kinds of environments. All of these problems are now being actively studied, and this chapter has been able to do no more than give a hint of the fascinating diversity of questions and answers which appear when we look at any living process.

## REFERENCES

The following list of references includes papers specifically referred to in this chapter, as well as additional reviews through which the student may find documentation of, and further information about, all the material presented in this chapter:


Beach, F. A. 1948. *Hormones and Behavior.* New York: Paul B. Hoeber.

———. 1951. Instinctive behavior: reproductive activities. *In:* S. S. Stevens, (ed.), *Handbook of Experimental Psychology,* pp. 387–434. New York: John Wiley & Sons.

Harris, G. W. 1955. *Neural Control of the Pituitary Gland.* London: Edward Arnold & Company.

Huston, T. M., and Nalbandov, A. V. 1953. Neurohumoral control of the pituitary in the fowl. *Endocrinology,* **52:** 149–56.

Koller, G. 1952. Der Nestbau der weissen Maus und seine hormonale Auslösung. *Verh. dtsch. zool. Ges.* (Freiburg), **1952:**160–68.

———. 1956. Hormonale und psychische Steuerung beim Nestbau weiser Mäuse. *Zool. Anz.* (Suppl.), **19** (*Verh. dtsch. zool. Ges.,* 1955): 123–32.

Lehrman, D. S. 1959. Hormonal responses to external stimuli in birds. *Ibis,* **101:** 478–96.

———. 1961. Hormonal regulation of parental behavior in birds and infrahuman mammals. *In:* W. C. Young (ed.), *Sex and Internal Secretions,* pp. 1268–1382. Baltimore: Williams and Wilkins.

Lehrman, D. S., Brody, P. N., and Wortis, R. P. 1961. The presence of the mate and of nesting material as stimuli for the development of incubation behavior and for gonadotropin secretion in the ring dove (*Streptopelia risoria*). *Endocrinology,* **68:** 507–16.

Parkes, A. S., and Bruce, H. M. 1961. Olfactory stimuli in mammalian reproduction. *Science,* **134:** 1049–1054.

Rosenblatt, J. S., and Aronson, L. R. 1958. The decline of sexual behavior in male cats after castration with special reference to the role of prior sexual experience. *Behaviour,* **12:** 285–338.

Valenstein, E. S., and Young, W. C. 1955. An experiential factor influencing the effectiveness of testosterone propionate in eliciting sexual behavior in male guinea pigs. *Endocrinology*, **56**: 173–77.

YOUNG, W. C. 1961. The mammalian ovary. *In:* W. C. YOUNG (ed.), *Sex and Internal Secretions*, pp. 449–96. Baltimore: Williams and Wilkins.

———. 1961. The hormones and mating behavior. *In:* W. C. YOUNG (ed.), *Sex and Internal Secretions*, pp. 1173–1239. Baltimore: Williams and Wilkins.

———. 1962. Patterning of sexual behavior. *In:* E. L. BLISS (ed.), *Roots of Behavior*, pp. 115–22. New York: Harper.

WILLIAM ETKIN

# 7

# Theories of Socialization and Communication

## Introduction

Our previous chapters have given us some insight into the ways in which social animal groups are organized, particularly in so far as that organization is determined by social status, territoriality, and reproductive activities. Except incidentally, they have not, however, enabled us to study the attractions which keep animals together in groups. We now wish to consider the concepts zoologists have developed about the nature of these attractive forces. For that purpose, we shall inquire into current ideas of the role, which innately determined behaviors and learning or experience play in the bonds that hold animals in groups.

For many years the concept of instinct as part of the behavioral equipment of animals has been in disfavor with many American psychologists. We need not go into the many reasons for this nor attempt to evaluate the position of those psychologists who still regard the concept as undesirable. It will be sufficient to say that, in this author's opinion, the concept has acquired new and significant meaning in the study of animal behavior, and its elaboration has provided insights into the nature of social behavior that are indispensable for modern views.

The present-day rehabilitation of the concept has resulted in large part from the penetrating observational and experimental work of some European zoologists, particularly Konrad Lorenz and Niko Tinbergen. In their writings these authors have expounded the value of comparative study of animal behavior under natural conditions or of laboratory study which utilizes methods and problems suggested by field observations. The name "ethology" has been given to this field to distinguish it from the more conventional animal laboratory studies of problems many of which have been suggested by human psychology. Our concern will be with only those aspects of ethological theory which help us to understand the social behavior of animals. We shall attempt to present this theory in its first, or classical, form as developed since 1930 (for summary, see Tinbergen, 1951; Lorenz, 1957) and shall refer only briefly to subsequent modifications by various ethologists. One reason for this procedure is that an understanding of present-day developments requires knowledge of the original theory. A second reason is that, as in so many other areas, research has gone off in many directions without producing a clear consensus of new understandings to replace those found wanting in the older theory.

Instincts will be viewed here as complex behaviors which function as adaptive units of action in an animal's life and which appear to be largely independent of learning. In other words, when we say of a behavioral pattern that it is an instinct, we assert that we believe that its fundamental elements are innately organized and require no specific type of experience as background for effective performance. Of course, the term has been used in other meanings in various philosophic interpretations of behavior, but we are here concerned only with the meaning that is operationally useful in experimental analysis. There is no a priori reason to expect that any activities as complex as the common instincts should necessarily be entirely uninfluenced by the previous experience of the animal. We find, in fact, that the interplay of experiential and innate factors in the development of behavior is very subtle and difficult to disentangle, a point repeatedly in evidence in the experimental chapters of this book.

By an instinct we also mean something that operates in the normal life of an animal as a means of accomplishing one of the activities required for its survival. In this respect, it is not a reflex, for a reflex is a bit of behavior that is not "goal-directed," in an adaptive sense, though it may, of course, be part of such behavior. Thus the withdrawal of a limb is a reflex which may be conditioned or unconditioned, but the setting up and defense of a territory appears to be an instinct, since it is a complex behavior functioning as a unit in the animal's life history and apparently not requiring specific experience with territories for its adequate expression in the adult.

The concept of "innateness" has been subjected to vigorous criticism by many workers, particularly Lehrman (1953) and Schneirla (1956). Without attempting to analyze this controversial facet of psychological theory, it would perhaps be well to clarify the use of terms in the present chapter. By "innate," in reference to either a structure or a behavior, we mean so regulated by hereditary factors as to require no specific conditions external to the organism for its development. Thus we say that eye color is innately determined in *Drosophila* and man. We are, of course, aware that there are an enormous number of interacting activities that intervene between the gene and its expression in the animal. These factors include the interrelations of various parts of the organism, particularly inductive, hormonal, and neurotrophic interactions. The general conditions of the environment also have to be favorable. In this sense, the eye primordia must have certain interactions with other parts of the organism and with generalized (nutritional, etc.) factors from the environment in order to develop successfully and express their "innate" characteristics. The word "innate," therefore, refers to a complex idea, not to a unitary factor or one that is entirely autonomous (cf. Lehrman, 1953, p. 343). For example, some cancers in mice are thought to be determined by innate factors, others are said to be acquired. In the former no special external condition is required; in the latter exposure to a specific virus or other agent is a necessary antecedent condition. Similarly, we shall speak of behaviors like leg-lifting by male dogs as innately determined. We imply by this that the place to look for the influencing factors is in the developmental physiology of the animal, including hormonal changes, etc. On the other hand, we think of retrieving behavior in the dog as learned behavior, and so if we wish to understand this behavior in a particular dog, we seek information on the special training that dog received prior to performing this particular act. In either case we could be mistaken about our assumptions, but the initial characterization is a useful first step, based on our unanalyzed experience with the behaviors in question.

The Lehrman-Schneirla criticism has served at least one very useful function. It has shown clearly that the adjective "innate" does not constitute an explanation. This criticism has become the focus of experimental analyses that have successfully shown the complex interaction of factors necessary for the development of certain behaviors. In showing the complexity of developmental and experiential interaction involved in behaviors formerly thought to be the simple and inevitable outcomes of genetic action, these experimental analyses support the contention of Lehrman and Schneirla—that the term "innate" is too vague to be useful. One of the valuable perspectives of ethology, however, is its view of behavior as part of the huge complex of animal adaptations to the environment. In this context "innate" seems very useful to the present author as a term to distinguish between those behaviors which seem to require and those which do not seem to require specific (learning) experiences for their expression. In that provisional sense we shall distinguish behaviors as innate or acquired. For example, we want to know whether the social behaviors described in earlier chapters, such as the selection and response to the appropriate mate in reproduction, require a special type of social experience on the part of the animal or can develop without any such experience. At least at the level of analysis of this chapter, the distinction between innate, instinctual patterns and acquired, or learned, patterns is useful. It is perhaps pertinent to remark that the difficulty in dealing with the concept of "innate" behaviors comes partly from the vagueness of the antithetic concept, "learned" behaviors, since not all alterations of behavior resulting from experience fit our concept of learning. For example, the strengthening of muscles with exercise or the increased response of an organ to a hormone or stimulus after previous exposure without reinforcement would not seem to constitute learning. The concept of learning is as difficult to delimit as that of instinct. In consequence, there has been some tendency to substitute the term "experiential factor" for that of "learning" and thus avoid rigidity in thinking about specific problems.

## The Ethological Concept of Instinct

### THE MOTOR COMPONENT

We may analyze an instinct in terms of the stimuli which evoke it and the nature of the responses to those stimuli. It will be convenient to consider first the responses, that is, the motor component. These are complex actions, such as pecking at particles on the ground in chicks or mating performance in rats. Since these tend to be highly uniform and recognizable as units, they have been called "fixed-action patterns." We have described some of these in Chapters 1 and 4.

According to ethological theory as originally developed by Lorenz and Tinbergen, the co-ordination of motor activities is accomplished in an organized part of the central nervous system called the "instinct center." It is important to note that such a center is not necessarily localized in one group of neurones or even in one region of the central nervous system. The concept is a physiological, not a morphological, one. Its important characteristic is that the parts are so linked together that once activated they provide for that appropriate sequence of muscular or glandular activities which constitute the behavior.

Such a concept of a center of organization is, of course, not particularly novel. The important aspect of ethological theory are the properties ascribed to the system. One of these is that in such a center, a specific type of energy is generated which tends to produce a pressure for discharge and without which the center does not discharge. The basis for this idea of action-specific energy are experiments indicating that the threshold for the response to a given stimulus is increased after draining of the energy by discharge of a center and is lowered by a period of rest which enables the energy to reaccumulate. This is a very generally observed phenomenon. It is seen, for example, in the decreased response of a rat to a female after copulation, the decreased hiding response of a bird to the stimulus of a predator, the cessation of nest building in a stickleback after a bout of nest building, or the failure of the flying-up response of a male grayling butterfly when repeatedly stimulated by models of a female (Tinbergen, 1951; Lorenz, 1957). Such decreases in vigor of responses can be shown not to depend upon fatigue of the motor organs, because these organs can be made to respond in connection with other behaviors without loss of effectiveness. Similarly, the sense organs can be shown not to be fatigued. Therefore, the change must be ascribed to some loss in the central nervous system. This loss is limited to the action in question, to a reduction of "drive" in that system. Therefore, it is inferred that the center for the organization of each particular type of instinctual behavior accumulates a limited supply of energy. This is depleted by repeated activation of the system. When none is left, the system cannot activate its characteristic behavior in spite of appropriate stimulation.

Another source of evidence for the concept of action-specific energy is the existence of so-called vacuum activity, that is, discharge of an instinctual action in the absence of any apparent stimulus. Lorenz (1957, p. 143) describes an instance of such behavior as follows:

I had once reared a young starling who performed the whole behavior patterns of a flyhunt from a vantage point *in vacuo,* with a wealth of detail that even I had, until then, regarded as purposive rather than instinctive. The starling flew up onto the head of a bronze statue in our living room and steadily searched the "sky" for flying insects, although there were none on the ceiling. Suddenly its whole behavior showed that it had sighted a flying prey. With head and eyes the bird made a motion as though following a flying insect with its gaze; its posture tautened; it took off, snapped, returned to its perch, and with its bill performed the sideways lashing, tossing motions with which many insectivorous birds slay their prey against whatever they happen to be sitting upon. Then the starling swallowed several times, whereupon its closely laid plumage loosened up somewhat, and there often ensued a quivering reflex, exactly as it does after real satiation. The bird's entire behavior, especially just before it took off, was so convincing, so deceptively like a normal process with survival value, that I climbed a chair not once, but many times, to check if some tiny insect had not after all escaped me. But there really were none.

Another important behavioral characteristic that may be understood in terms of action-specific energy is that of displacement activity. It appears that when an instinctual activity is aroused strongly but its expression is at the same time inhibited either by simultaneous activation of another and incompatible instinct or by failure of appropriate releasing stimuli to appear, the animal may exhibit a different behavior pattern and one irrelevant to the situation. For example, fighting roosters when evenly matched often show evidence that fear is inhibiting their aggressive tendencies; the incompatible responses of flight and fight are both aroused. In such a case the animals will often break off their fighting and peck at the ground as though suddenly interested in feeding on non-existent grain. According to the theory, we may suppose that neither the aggressive centers nor the flight centers were able to discharge their action-specific energy sufficiently and the energy from these centers "spilled over" or "sparked over" to the feeding center. When a courting bird (for example, the yellow-eyed penguin) executes a vigorous display but the intended mate fails to respond appropriately, the animal will often start preening itself. In this case, we may suppose that mating energy "sparked over" to the preening center. Such irrelevant behaviors are called displacement behaviors. They are behaviors activated not by their own, or autochthonous, centers but by energy derived from other, or allochthonous, centers. As we shall see subsequently, this concept of displacement activities plays a considerable role in the ethological interpretation of social behavior. Here we need only note that it is a type of activity suggesting that action-specific energy can be accumulated in instinct centers. In Chapter 8 we will see that Tinbergen and other ethologists now favor an interpretation of the origin of some displacement behaviors somewhat different from that one given above and based on the reduction of inhibition.

Related to displacement activity and similarly suggesting the presence of action-specific energy is the common tendency, described in Chapter 1, for a subordinate animal to pass along punishment it has received from its superior to a subordinate in the social hierarchy. Since such action is behavior-specific (autochthonous), it cannot be said to be a displacement activity in the technical sense, although of course the action is "displaced" to a new object. This passing along of an instinctive act to an object other than the one presenting the optimal stimulus has been called redirection. It occurs commonly with respect to sexual activity, parental care, and even feeding reactions. It, too, suggests the accumulation of action-specific energy with the lowering of thresholds for release of motor centers.

Evidence from many other sources may be adduced to support the concept of action-specific energy. For example, the development of a type of instinctive behavior in the life of an individual or its arousal in the course of the yearly cycle of behavior often shows evidence of the gradual development of a particular "mood" for such behavior (see Chaps. 1, 4, 5, and 6 for examples of the development of such behaviors). The incipient movements characteristic of early discharge phases of a response are called "intention" movements. Thus we spoke earlier of territory establishment in song sparrows being shown briefly and in a desultory manner early in the season. Gradually such behavior is displayed with more and more vigor and in a more and more sustained manner. The intention movements pass over into the fully expressed instinct. Similarly, within the territory, the settling down to a final nest site is seen in many birds to follow a similar period of increasing intensity of exploration of nest sites and handling of nesting material. Intention movements are, of course, also shown in adults. Before flight, many large birds show flying-off intention movements. Such intention movements enter into social signaling in an interesting way which will be discussed later.

It is also noteworthy that physiological investigation of the nervous system gives indication of neurological centers from which some complex activities are controlled as units. As explained in Chapter 2, certain areas of the hypothalamus have been found to act as stimulative or inhibitive centers of organized rage, sleep, or feeding behaviors (Fig. 7.1). Presumably these are principal parts of the apparatus reg-

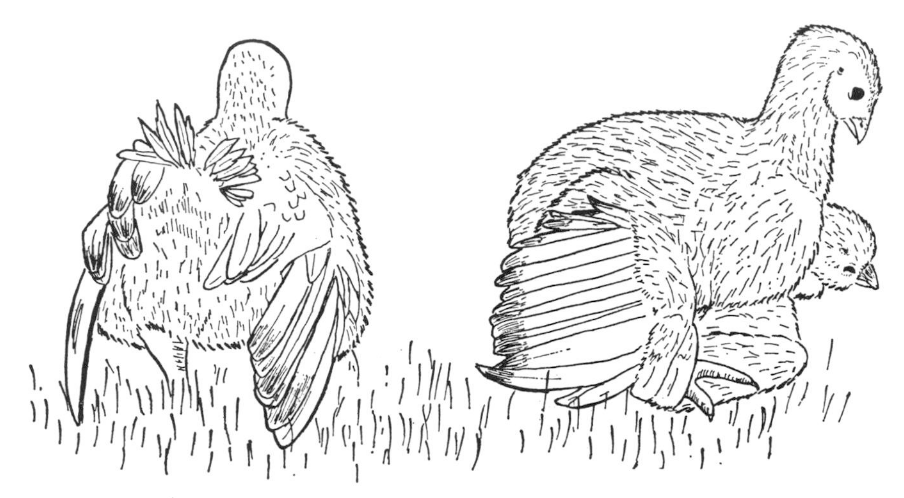

**FIG. 7.1. Courting Behavior in Androgen-Injected Young Turkeys** Immature turkeys injected with male hormone carry through courting behavior in a highly organized fashion without previous experience. Note that the animal on the left is strutting with wings lowered and tail fan (as yet undeveloped) spread and turned toward the stimulus object (a mount of a squatting turkey). The bird on the right is correctly oriented and is copulating. The pattern of behavior is thus seen to be already organized and ready for action in the immature bird. (After M. W. Schein and E. B. Hale, *Animal Behaviour,* 7 [1959].)

ulating instincts which include these behaviors. Other neurological studies have shown that certain regions of the central nervous system maintain rhythmic discharges even when isolated. Respiratory and locomotor regions of fish and amphibian nervous systems show such self-maintained rhythmic activity in spite of apparent isolation from sensory impulses. This physiological evidence of autonomous action in neurological centers applies to the expression of relatively simple activities that, at most, contribute to only the motor aspect of some instincts. Perhaps the analogy to ethological concepts should not be pushed too far. The centers conceived by ethologists involve, as we shall see, at least additional mechanisms for the selection of stimuli (Tinbergen, 1951).

We have referred to typical instincts as units of behavior because, though composed of many individual acts, the behavior tends to hang together as a unit. Many such behaviors, however, seem to fall into two subdivisions; the first has been called appetitive behavior and refers to the generalized seeking or exploratory behavior that an animal shows when an instinctual mood is first aroused in it. The stickleback, as the reproductive season approaches, shows a wandering behavior accompanied by an exploration of nooks and crannies along the water's edge. We might say he is "looking" for a suitable territory. Similarly, the starling's search of the ceiling, described by Lorenz in the quotation above, constitutes appetitive behavior. Wallace Craig (1918), whose early studies first suggested this concept, wrote: "An appetite so far as externally observable is a state of agitation which continues so long as a certain stimulus is absent." Such behavior is strikingly variable in expression and emphatically gives the appearance of being "purposive," in that it continues indefinitely until what might appear as the "goal" has been realized. In fish, bird, or frog, searching continues restlessly until achievement of the goal brings it to an end. Thus the hawk explores and maneuvers until the chance to pounce on prey appears, and the frog turns to follow a crawling creature until the stimulus to flick his tongue in capture appears.

The consummatory act, on the other hand, is quite stereotyped and appropriately described as a fixed-action pattern. The flick of the frog's tongue is so simple there would not seem to be much occasion for variation. The same fixity applies, however, to rather complex actions, such as the food handling of the starling described above or the territory-defense behaviors in the fish or birds as we discussed it in Chapter 1. Some authors, Lorenz, for one, have suggested restricting the term instinctive behavior to these stereotyped consummatory acts. Indeed, if we are to insist that instinct cannot have any element of learning in it, this might seem appropriate; for, as we shall see later, learning enters conspicuously into appetitive behavior. However, the exclusion of the appetitive behavior from the instinct destroys the unity of the activity and makes it difficult to visualize the role of the activity in the life of the animal. From the point of view of the animal's survival, the appetitive behavior provides the seeking of necessary goals, and the consummatory behavior represents the achievement of the goal. An important reason for considering both as part of an instinct is that, in this concept, both derive their energy or drive from the same central mechanism. When that mechanism is depleted, both behaviors disappear.

From its nature it is obvious that appetitive behavior does not serve to discharge the action-specific energy characteristic of an activity for if it did the appetitive behavior would cease. It is the consummatory act which discharges action-specific energy. Thus, after the consummatory act is carried out, the appetitive behavior ceases, at least for a while, until the energy store accumulates again. After discharge, the animal appears "satiated" with respect to that behavior. This leads to the concept that the "goal" of the appetitive behavior is the expression of the consummatory act and not, as we are tempted to think, the satisfaction of a physiological need such as hunger, rest, etc. In other words, what the hungry hawk described above is "seeking" is the opportunity to pounce upon prey, not the food as such. The activity of pouncing depletes the instinct center and thus quiets the restless seeking behavior. Of course, the food that results from the successful hunt under natural conditions stills the hunger of the animal. In the absence of hunger the action-specific energy is not so quickly re-formed in the center, and consequently the physiologically satisfied animal does not return to the appetitive behavior of hunting as quickly as the unsuccessful animal. The satisfaction of the physiological need does play a role in the behavior, albeit a secondary one.

Evidence for this concept of the consummatory act as goal can be seen even in natural conditions. For example, the young passerine bird in the nest shows gaping activity; that is, it opens its mouth wide and stretches its neck upward whenever a parent lands on the edge of the nest. Ordinarily the bird gets fed at this time, but if a nest parasite such as the cowbird or cookoo is present, the parasite may get all the food. The gaping response of the other birds in the nest nonetheless diminishes considerably in the same way as if they were fed. In dogs, too, it has been found that the drinking action of a thirsty dog ceases after it drinks a certain amount. This occurs in a normal dog and in one in which a pipe has been inserted into the esophagus so that all the water swallowed escapes from the animal, and its thirst cannot be said to have been satisfied physiologically. Of course, the action-specific energy of the system is quickly re-formed in the deprived animal, whereas the sated one may not again be ready to discharge the consummatory act so soon. That the "goal" of an animal's appetitive behavior is the consummatory act, rather than the gratification of a specific "need," is one of the basic tenets of the modern concept of instinct. It is fundamentally this conceptualization which frees us from the mysticism that has clung to the older ideas of instinct. The notion that the goal of instinctual behavior is a physiological need inevitably implies purposefulness and rationality, even consciousness on the part of the animal. Even where objective methods of experimentation are employed, the true nature of the goal may be missed. As Tinbergen (1951, p. 106) says, "Even psychologists who have watched hundreds of rats running a maze rarely realize that, strictly speaking, it is not the litter or the food the animal is striving toward, but the performance itself of the maternal activities or eating." Perhaps Tinbergen should have gone further and pointed out that the maze-running activity itself is part of a territorial behavior that is "satisfied" only by the activity of exploring the maze. The drive to explore its environment is a goal of behavior in many animals, irrespective of any "material" reward that such exploration produces.

### THE SENSORY COMPONENT OF INSTINCT

The ethological concept of instinct suggests that the neural energy for instinctual action is bound up in a neural center and the path of discharge laid out in advance; the only thing lacking is a stimulus from the environment that determines the timing of the discharge. From this point of view, the concept of the discharge of instinctual acts by simple sign stimuli or signals does not seem strange or unexpected. Yet it is this concept, more perhaps than any other, that is a basic novelty of the system proposed by the ethologists.

According to this concept, the discharge of each consummatory act is blocked by a mechanism which can be released only by stimuli reaching it from the animal's environment. The blocking mechanism has been called the internal releasing mechanism (IRM). The idea of release is emphasized in this terminology because, from the point of view of behavior, it is the release rather than the blocking that requires explanation. The stimuli from the environment that release most IRM's that have been studied are found to be simple and specific (Fig. 7.2). Each stimulus fits its IRM as a key fits a lock. That is to say, it contains characteristics which make it distinctive, characteristics which do not ordinarily occur in the environment except in the appropriate releasing object. Thus the mating response of an animal should be given only to an appropriate mate; the feeding response should be given only to appropriate food. Rationally considered, appropriate food for a frog might be said to be any organic material not specifically harmful. Experiment shows, however, that the frog snaps at any moving object of small size but not at stationary food. Consideration of the normal life of a frog shows, furthermore, that practically the only objects of appropriate size that move in its environment are small live organisms. These are, of course, appropriate food. As a practical matter, therefore, it is advantageous for a frog to

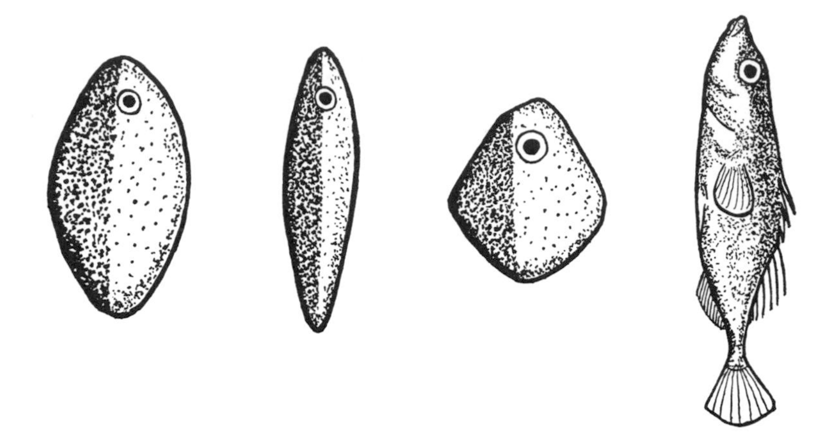

FIG. 7.2. Experimental models used in testing for the releaser of territorial defense behavior in the three-spined stickleback are shown here. The right figure is an accurate representation of a male, but the belly is light colored in contrast to that of the normal breeding male, which is red. The other models depart to varying degrees from the normal appearance. However, the ventral sides are colored red. Each of these elicited much more attack from the territorial male than did the right figure. This experiment shows that it is the distinctive red belly rather than other characteristics of the rival which serves to release the fish's attack behavior. (After N. Tinbergen, *Study of Instinct,* 1951.)

snap at any such object, for the combination of movement and small size rarely occurs in inappropriate objects under normal circumstances. Of course, in the laboratory, we may wonder at the stupidity of a frog starving "complacently" to death in spite of the fact that pieces of fresh meat are put into its cage every day. The releaser for the instinctual behavior here represents about as simple a combination of characteristics as would practically "key-out" the normal nature of its food in natural conditions.

It is the contention of this theory that the sign stimuli or signals that release instinctual behaviors are combinations of characters, or even single characteristics, that specifically identify the appropriate object and do not occur ordinarily in inappropriate objects. The reaction is given to these characters and not to the situation as a whole. To quote another example from the feeding mechanisms of animals (for the moment we wish to avoid social responses), we may point to the striking reaction of a rattlesnake in the dark to any small, warm object that moves in its neighborhood. To a rattlesnake in its normal habitat this can only be a small mammal, which of course is appropriate food. In the laboratory a warm electric bulb wrapped in cloth elicits the discharge of the feeding mechanism just as effectively as proper food. The combination found to release the feeding mechanism of a parasitic tick—in this case, the drilling action of the proboscis—was a combination of warmth and butyric acid odor, a smell common to the skin of many mammals. The tick may remain quiescent on a bush for months or even years from one meal to another until a warm object smelling of butyric acid brushes by. The tick then releases its hold, drops on the object, and inserts its feeding month-parts (Lorenz, 1957). This again is the rare combination of simple stimuli that occurs in the normal tick's environment only in association with a passing mammal, the appropriate object for feeding for the tick. Specificity and simplicity are almost perfectly combined. Only the evil genius of the experimenter would smear butyric acid on a rock warmed in the sun to deceive a tick into ruining its proboscis.

There are a number of complications to the idea of the consummatory action which deserve some consideration. One is the fact that many actions which appear to be fairly simple have been shown to be composed of a series of separate actions. For example, the approaches of various insects to food or sex objects can be separated into at least two reactions in a chain. Bees and some moths are attracted to flowers by odor, color, and details of form, each stimulus being effective at a different distance (Baerends, 1950). Some male moths find the female by a response to odor from the female's scent glands which activates a reaction system. This induces the animal to fly upwind, a response which would, of course, normally bring it into the vicinity of the female giving off the scent. Some of the instances of sign stimuli may appear complex because a sequence of separate responses, each with its own appropriate releaser, may be mistaken for a single response.

Another elaboration of the concept of instinct, at one time emphasized by Tinbergen, is that instinctual behavior is organized into a hierarchy. The top levels of the hierarchy govern the more general activities. Energy released from them activates lower levels, governing more specific activities. For example, the top level of reproductive behavior in a stickleback fish sets up the appetitive behavior of inshore migration and searching for a territory. The releasers which act on this top center are presumed to be those which determine the localization of a territory. Though they are unknown in detail, they presumably include weeds and warmth. Once the territorial center has been activated by energy released from the general reproductive center, the animal shows territory-guarding behavior. This includes the appetitive behavior of searching through the area, especially patrolling near the boundaries. Should a male presenting a red belly swim into the territory, this would release fighting behavior; or contrariwise, should a female with a swollen white belly appear, courting behavior would be released. The release of the lower centers initiates the consummatory acts, which drain action-specific energy from their centers. Although this concept of a hierarchical organization is theoretically interesting, it does not seem to have been particularly helpful in experimental analysis and has been de-emphasized in recent ethological writings.

## The Stimulus in Learned and Instinctual Reactions

The essential difference between the ethological concept of the stimuli of instinctual actions and those of learned reactions is the relative simplicity and fixity of the former. Lorenz (1950) accepts the concept that in learned reactions the stimulus is a *Gestalt,* or a whole. Such a stimulus is effective even if considerable detail is altered, as long as the over-all unity is maintained. For example, if an animal is trained to respond to a circle, its response is not disrupted if, instead of a continuous circle, a dotted outline is presented or if the axes of the figure are altered, within limits, to produce an ellipse. According to this notion, the perception upon which learning depends is the perception of a field, or whole, which may be given by a physically imperfect representation. In contrast to this, sign stimuli are characterized by the dependence upon single, or a few, aspects of the stimulus without regard to the loss of any resemblance to the wholeness of the normal stimulus. Thus the territorial defense behavior of the English robin is set off by the posing display of a rival. If this response were to a *gestalt* perception, we would expect that a dummy would serve to set off a response proportional to the over-all similarity of the dummy to a natural robin. Yet experiment reveals exactly the opposite. One can remove the head, the tail, the back feathers—in fact everything that makes it look like a robin (to the human observer)—and still elicit a good response as long as a bunch of red feathers is retained. On the other hand, if a stuffed, complete robin is presented but with the red breast dyed brown, no defense reactions are elicited from the territory owner. To our eye this model closely resembles a robin, but to the robin's releaser mechanism a simple bunch of red feathers on a wire is much more effective (Lack, 1943).

We need not attempt here to go into the theory of perception. It is obvious that some configurational characteristics of a limited kind may play a role in releaser perception, as indeed Tinbergen (1953) finds in the escape response of fowl to the outline of a flying hawk, for example. Here it is not only the shape of the figure that releases the escape response but the shape in relation to the direction of movement. Even if we accept the observations of Tinbergen in this regard (they have been questioned by other experimenters), it is obvious that we are dealing here with a simple perception which ignores much of the characteristics of the stimulus object. In any event the relative simplicity of releasers and the absence of strong *gestalt* tendencies in perception are outstanding characteristics of instinct releasers. We shall see that these aspects are of great significance for their role in social communication.

Different elements of the releaser have their separate effects, so that they summate and together are more effective than each aspect is by itself. This phenomenon of summation is discussed more extensively by Tinbergen from the point of view of the evolution of signaling systems in the next chapter. We may note further that the naturally occurring releasers may not be maximal in effectiveness with regard to each component characteristic. As a consequence, it is possible to provide artificial stimuli that are even more effective than the natural ones. Thus certain birds, when given the opportunity, will choose to incubate monstrously large or excessively spotted eggs in preference to their own. These so-called supernormal stimuli give striking evidence of the mechanical nature of the animal's responses to releasers.

## Imprinting

We have not said anything hitherto with regard to the question of whether the stimulus recognition is innate or learned. It was tacitly assumed that the organism, somewhere between the sense organs and the IRM, has an innate capacity for recognizing appropriate sign stimuli. It would be difficult to see how such a sign stimulus as a red breast could be learned, since the young robin does not see the reaction performed before it is itself capable of carrying it out—in the first springtime of its life. An element of learning does enter into stimulus recognition of some IRM's. This was discovered in experiences with incubator-hatched geese and other fowl. It was found that if the young birds are handled by the experimenter during their first few hours of life, they will thereafter react to him and to other human beings as they normally would to their parents. In geese, for example, the young will faithfully follow the experimenter wherever he goes. They will seek the protection of his body whenever they hear the normal warning cry of the species. This phenomenon of the learning of a releaser stimulus in a short exposure to it has been called imprinting (Lorenz, 1957). It was discussed briefly in Chapter 4 as an aspect of parental care.

Imprinting has many characteristics which distinguish it from ordinary learning processes. In fact, Lorenz has often insisted that it cannot be considered to be learning. Since, however, imprinting is clearly a behavioral effect of experience, it seems arbitrary to distinguish it from other learning. We shall consider it a type of learning and note the differences from other kinds. One of these distinguishing characteristics is that the capacity for it is limited to a brief, specific "critical period" of an animal's life. If imprinting is not accomplished during the first few days of a hatchling's existence, it will not "take" at any other time. Such an animal will fail to respond appropriately to any other animal, and no amount of association with members of its species will bring out the response of following. Chicks have been found to imprint to one another as well as to the hen in the first seventy-two hours of their lives (Guiton, 1959). Hess (1962) finds that a principal factor in determining the strength of imprinting is the effort expended by the young in the process. He also emphasizes the differences from ordinary reward-conditioned learning (Fig. 7.3). Most remarkable, indeed, is the fact that when the animal which has been imprinted to a human being becomes mature many months later, it will show courting responses to humans even in preference to its own species. It is thus clear that imprinting can take place long before the reaction with which it is associated matures. This is a phenomenon often observed in birds. Hand-raised birds commonly display courting behaviors to the trainer's hands (Schein and Hale, 1959).

Not all the instinctive reactions of a given organism are directed to the same imprinted object. For example, Lorenz (1957) describes how jackdaws imprinted to himself would court him but would show

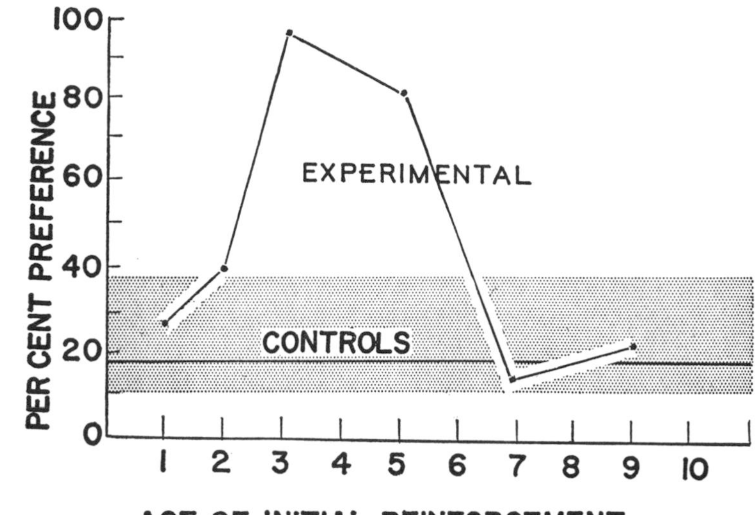


FIG. 7.3. When inexperienced chicks are presented with a choice of a white triangle on a green background or a white circle on a blue background, they show a small preference for pecking at circles. The triangles were selected only about 19 per cent of the time in tests of about four hundred chicks. Groups of chicks were exposed to these stimuli at various ages from one to ten days, and their pecks at the triangles were reinforced by food rewards. When tested later against these same stimuli the effect of their early experience was shown only by animals receiving reinforced experience on the third and fifth days. This demonstrates the existence of a critical period at three to five days for learning the pecking response. (After E. Hess, E. Bliss [ed.], *Roots of Behavior,* 1962.)

flying-off behavior only to crows and would attack him as a predator when he held a black waving object, the sign stimulus equivalent to a struggling jackdaw, in his hand. Each reaction had its own releaser. Under natural conditions the innate releasers (for predator attack) and those dependent upon imprinting (for courting) would both be given to jackdaws only, since an animal raised by its parents would be so imprinted.

Another characteristic of the imprinted releaser is that it is of a general, supra-individual character. That is, imprinting is to the human or other species, not to one individual human being. Again this is most strikingly shown in courting behavior. Instances have been recorded of male turkeys imprinted to a male caretaker, not only showing courting behavior to him, but showing rivalry and antagonistic behavior to women, presumably because the loose clothing of the women suggested male turkey characteristics, such as the wattles. This long-term and very stable characteristic of imprinted stimuli is of utmost importance, because, as we shall see, it affords a possible explanation of many aspects of socialization in animals. It should be noted, however, that the generalized bond established by imprinting may be refined and restricted to one individual by later learning processes. Thus, though ducklings at first follow any duck, they normally learn soon to restrict their following response to their own mothers.

## Specificity of Learning Processes

It is not our intention here to attempt to consider the nature of animal learning generally but simply to confine our discussion to certain specific aspects that are more or less directly applicable to the problem of socialization of animals. One of these is that animal learning often shows a high degree of specialization. It thus departs from the common conception of learning ability as a generalized phenomenon, a single capacity which may be called intelligence. We usually assume, further, that there is a clear-cut correlation between the degree of development of the brain and the various aspects of intelligence, such as memory, rate of learning, complexity of learning, and insight learning. In a broad phylogenetic view, this concept may indeed be justified, for it would certainly appear that annelid worms are inferior to fish in any reasonable test and, similarly, that mammals are correspondingly superior to amphibians. The nervous system has made such enormous strides between these grades of animal organization that the variations to be found within each group and in response to different tests may be ignored in a broad comparison between the stages. However, when we ask specific questions concerning learning capacity, we are immediately led to realize that there is another factor, namely, ecological adaptation, which plays a very important role and which may override the broad phylogenetic progression in mental capacity.

If, for example, we ask about memory, we not only find that some birds such as parrots, are credited (on reliable evidence) with memory capacities for individual recognition extending over years and comparable to the highest ability shown by mammals, but we find evidence of something comparable in fish. For example, salmon are known to be able to recognize in some way the stream in which they were hatched. They return to it for spawning after five years or more of life in the open ocean. Is this ordinary (conditioned) learning or imprinting? Does it depend upon a memory trace (engram) in the nervous system? We do not know, but it would appear highly arbitrary to deny that it is memory of a kind. If we consider, as another example, the learning of localities and pathways, as in maze learning, we find great discrepancies between phylogenetic predictions that capacities increase with size of brain and actual performance. Wasps, whose small nervous systems must be morphologically very simple compared to that of vertebrates, nevertheless show an unexceeded ability to learn a homing response. Various digger wasps, for example, after excavating a burrow, cover it carefully, completely concealing the opening. They then circle over it once or twice and are off on a hunting expedition to stock their nest holes. They may not return for hours or even days and yet have no difficulty in returning to the exact spot. Experiments show that this capacity depends upon the learning of visual cues furnished by neighboring landmarks. Such learning appears to be achieved in a brief orientation flight and to persist for days. The human observer often confesses his inability to equal this achievement (Thorpe, 1956). Since song birds tend to return to the neighborhood of their birth and to the very territory held the previous year, it is obvious that they too retain a memory for the necessary details, not only of the topography of the territory itself, but of the surrounding country as well. It is clear, of course, that these "mental" capacities are part of the adaptation of the animal to its particular mode of life and do not imply a general long memory or capacity for learning complex patterns other than those involved in the particular functional activity. Indeed, it may be that many other of the learning processes commonly studied in the laboratory by experimentalists also represent similar isolates and therefore give entirely erroneous impressions of the mental capacities of the subjects. For example, rats have been used very extensively in maze-learning tests. It is known that the rat's ability for maze learning is about the same as a human's ability to learn a pencil maze when blindfolded. Should this be taken as a general measure of learning capacity, even of capacity of one general kind? Rats under natural conditions have a tremendous exploratory drive, which leads them to explore every new aspect of their territory. Under such conditions rats come to know their territories very thoroughly. The biological usefulness of this knowledge becomes apparent when a predator, such as an owl, appears. The rodent then shows a capacity for dashing for the nearest hiding place without the least hesitancy or circumambulation, irrespective of its position at the moment. Knowledge of the topography of their home range is thus of overwhelming importance for the survival of rats. The high capacity for maze learning is seen, therefore, to represent a specialized ability of the rat and can hardly be expected to bear any close relation to the general capacities of its level of nervous organization. Incidentally, it may be noted that what psychologists call "latent" learning of mazes —that is, learning which takes place when rats are simply left in a maze and permitted to explore without being given a reward—appears, from the ecologists viewpoint, to be learning that is highly rewarded. It is obvious that knowledge of its environment is one of the primary "needs" of a rat, and the achievement of that knowledge by exploration thus satisfies just as food satisfies hunger.

The biologist likewise tends to take a skeptical view of the phylogenetic interpretation of many other kinds of comparative learning capacities which are not considered in relation to the animal's mode of living. For example, psychological experimentation indicates that lower animals, even apes, have very limited capacities for carrying through delayed responses, a few minutes being top score in the usual laboratory test. In the laboratory situation the facts are clear enough. Yet, when one considers that many animals, birds as well as mammals, carry through activities involving long delays such as returning to the site of a kill or buried or hidden food, it is difficult to believe that short delays really represent the limits of the mental capacities of animals (Thorpe, 1956). One animal, such as the macaque monkey, may be particularly good in a type of problem such as the "Umweg" problem, that is, a problem requiring that the subject go in a roundabout way to the solution. This capacity may indicate only that that animal normally lives in an environment—amid trees, for instance—where the roundabout path is frequently the only solution to naturally occurring problems of obtaining food. Therefore, monkeys do well in such problems, and dogs do not. It may be that this difference is related more to their ecology than to the evolutionary status of their nervous systems.

This characteristic of animals, whereby their abilities for the learning of particular types of responses are related to their ecological requirements, is one of the fundamental points of emphasis of ethological studies. Tinbergen (1951) refers to it as a function of differences in the "innate disposition to learn." This specialized character of animal learning illustrates the fact that learning ability often depends on special innate factors, just as the expression of instincts frequently depends upon special factors of experience. Thus, an animal's behavioral characteristics do not segregate themselves strictly into innate or learned categories. Related animals with similar brain structures may differ considerably in their innate disposition to learn particular items (Fig. 7.4).

Our conclusion from the foregoing discussion is that mental capacities of learning and related phenomena correlate in details better with ecology then with phylogenetic status. From the point of view of social behavior, this concept is important. It indicates to us that specific items of learning, such as individual recognition, locality and homing learning, memory for early experiences, etc., can be surprisingly complex even in animals which appear to be otherwise very limited in their mental capacities. Thus we cannot, on phylogenetic grounds, consider that a phenomenon cannot be learned in a particular animal because it is too complex for the animal's grade of organization. The common phylogenetic generalizations about learning, such as that insects and birds are largely animals of instinct, whereas mammals show much more learning in their behavior, cannot be applied a priori to any particular activity with confidence. Indeed, Hinde (1961) has questioned the verity of the common opinion of the superior learning capacity of the mammalian, as compared to the bird brain.

## Motor Patterns in Learning

Much learning leaves motor patterns unmodified but simply associates such patterns with new environmental stimuli. It is, for example, obvious from what was said above about maze learning in rats that maze exploration is a motor pattern that does not need to be taught to rats but comes to them "naturally," presumably innately. Certain other animals, such as chickens, have been used in maze-learning experiments but without much success, because maze running is not a normal part of their mode of life. Sometimes the experimenter went so far as to use electric shock or blasts of air to move the animal along in the maze. Hediger (1955), in his discussions of the training of circus animals, emphasizes the use of natural capacities. Sea lions, for example, hardly need any teaching to balance balls on their snouts; they will of their own accord play with floating pieces of wood by tossing and catching. Apparently, flexibility of neck correlated with

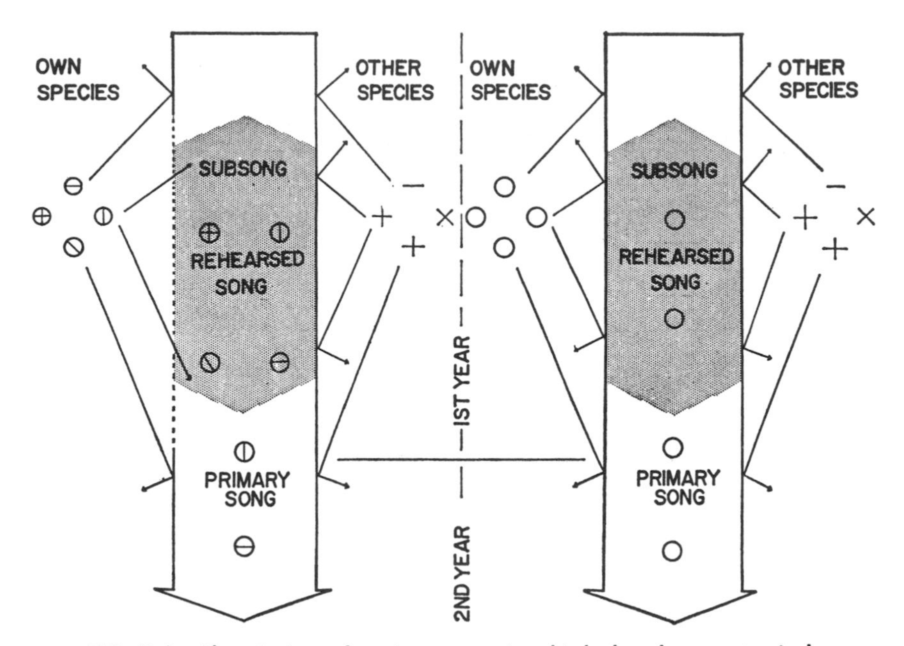


FIG. 7.4. The singing of various passerine birds has been extensively studied in birds raised in isolation and variously exposed to the song of their own or other species. During the first year the birds show some singing and call notes but not the distinctive song of the species. This is called subsong. It is followed, usually by the fall of the first year, with the introduction of some motifs of the characteristic song. The following spring, the definitive song of the species, the primary song, is produced, usually after a brief period during which subsong and rehearsed song are recapitulated. In some species, as the corn bunting, the evidence indicates that this developmental process proceeds effectively, independently of any learning from other birds. This is indicated in the *right diagram* where the arrows from the symbols for motifs of other species or of its own species are shown as reflected without effect on the developmental process. This condition is regarded as probably the primitive one in song birds. Most present day birds such as the blackbird, white throat, canary, and chaffinch, however, show a period of susceptibility to the influence of other members of their own species. In this case, the motifs of other species are shown as rejected, but those of their own species as penetrating and having an effect during the susceptible period (*left diagram*). In those birds refinements of the motifs are learned from other members of the species, and the young of these species complete their full repertoire of primary song by learning. Some species as the red-backed shrike are able to add motifs from other species and thus are true mimics. (Modified after Lanyon, *in* W. E. Lanyon and W. Tavolga (eds.), *Animal Sounds and Communication,* 1960.)

FIG. 7.5. Some birds, such as goldfinches and jays (*above*), readily learn to secure food attached to strings by pulling up the strings with their feet. Others, such as robins and wrens, do not manipulate the strings and do not learn the trick. To an important extent the ability to solve this problem depends upon innate motor patterns for manipulating objects with the feet. In some cases these are so well developed that very little experience seems necessary and the learning achievement takes on the character of "insight." In others previous experience is an important factor in the animal's ability in learning this trick. (After W. H. Thorpe, *Learning and Instinct in Animals*, 1956.)

precise head movements controlled by binocular vision is part of their equipment for hunting fish. They tend to use these capacities under other conditions. Training of animals proceeds most successfully when it confines itself to using the motor equipment used by the animal in its normal mode of living. When we speak of a reaction as learned, we may be referring only to one aspect of it—the association between stimulus and response—but the response itself may have an innately determined co-ordination of parts (Fig. 7.5). This is not to deny that genuinely new patterns of motor co-ordination may also be learned. We wish simply to emphasize the other side of the coin discussed before. Just as previously we stressed the idea that the fact that a response appears to be innate does not preclude the existence of elements in it that depend upon experience, so we wish now to point up the fact that because a response is commonly considered learned, this does not mean that large elements of it are not determined or restricted by innate factors.

## Social Releasers and Kumpan Theory

The ethological theory of instinct has given us a new insight into the nature of animal socialization and society. The basic concept here is that animals carry in their own structure and behavior the releasers that evoke appropriate social responses in their fellow group members. Such releasers are called social releasers. As an example of this, we may mention the fact discussed in Chapter 1—that, in territorial animals, males arouse antagonism of other territory-holding males whenever they enter their domain. Experimental analysis of the way in which the territory holder recognizes his rival has shown that this is dependent in many birds and fishes upon relatively simple signals. For example, in the English robin, the red breast is the primary sign stimulus for the attack (Fig. 7.6). The animal thus bears on its own body the signal to release the appropriate social response of its rival.

Simple morphological sign stimuli are usually supplemented by peculiarities of behavior. Some of these, such as the posturing and singing from high perches by song birds, may be considered as ways of displaying the animal's distinctive morphological and auditory characteristics. Others, such as bill-clappering of cranes or the deep bow followed by hissing and snapping of herons, may not be dependent upon special physical characters for their effectiveness. At any rate, all of such structures and behaviors serve to elicit appropriate behaviors in another member of the species. If this be a male, they may serve to intimidate him to leave or, contrariwise, arouse him to fight for the territory. The female's response to the same signals may be quite different, at least, if she has reached an appropriately advanced stage in sexual development. She may respond by passive but insistent staying in the territory, in spite of the attacks of the male. Here we see another characteristic of social releasers. They act in situation-specific situations and are responded to according to the respondent's condition or mood. Thus the same sign stimuli may arouse territory fight, flight or courtship behavior, or, from a young bird, no particular response at all. Two animals, which may be distinguished as the "sender" and the "receiver," in analogy to radio transmission of signals, stand in one particular relation to each other

FIG. 7.6. Threat display of a robin redbreast (*right*) at a stuffed specimen (*left*) mounted in his territory. The red breast which is the releasing signal is made conspicuous by this posture. (After D. Lack, *The Life of the Robin,* 1943.)

with regard to a particular sign stimulus. This relation may be the mate-relation or rival-relation. They may be first one, then another; a female first unready for mating may respond as a rival and later as a mate. As human beings may be companions to one another in bowling or in business without being related in other ways, so animal companions, when their relation depends upon sign stimuli, bear only situation-specific relations to one another. The German students of behavior use the German term "kumpan" for such a situation-specific companion. The word in German apparently more nearly carries that connotation than does the English equivalent. It would, therefore, be convenient to retain the term for a situation-specific companion. It is at once technically accurate and satisfyingly familiar for our purpose.[1] The fundamental role of releasers in social behavior then may be summarized as follows: Animal kumpans furnish to one another the appropriate sign stimuli to release the behaviors adequate for a particular social interaction between them. These interactions are

[1] Anglicized plural—kumpans.

specific: male for female, rival for rival, parent for child, flight companion for flight companion, etc. Unlike learned reactions which involve the recognition of the individuality of each animal as in the dominance hierarchy, the kumpan functions merely as the bearer of the releaser for a particular behavior and is not recognized as an individual (Lorenz, 1950, 1957; Tinbergen, 1951, 1953).

## Evolutionary Trends in Social Releasers

One of the most significant aspects of social releasers and one in which they differ from non-social releasers is that, being parts of organisms, they are subject to evolutionary change (Fig. 7.7). The releaser mechanisms are subject to selection as are other characteristics of the organisms. There is, therefore, a tendency for them to evolve into more and more effective types. This evolution proceeds in both members *pari passu*, for the "sender" and the "receiver" must always be in tune with each other. This lock-step evolutionary process permits near perfection to be attained and accounts for many of the extraordinary specializations and varieties of pattern, color, and sound that animals have evolved as signals.

Effectiveness implies a number of characteristics. One of these is distinctiveness. That is, the releaser will be effective in so far as it is readily and clearly differentiated from those of other species or other reactions in the same species. If it is distinctive, it will not evoke its action in an inappropriate kumpan. For this reason, we find that movements and structures serving as releasers tend to diverge from one another in evolution much as ordinary structures do. From a common, original type they tend to become separated into a variety of specialized forms that distinguish each species from the others. Thus each species of song bird has a distinctive song or other display character that prevents confusion between species. These evolutionary trends in signals are discussed in detail in Chapter 8.

Besides distinctiveness, another evolutionary trend to be found in social releasers is that of developing conspicuousness (Fig. 7.8). We have seen many examples of this with respect to courtship behaviors, such as those of the peacock; although in these cases we inferred that physiological co-ordination of mates was also involved (Chaps. 4 and 6). Simpler examples of releaser function are found in the many characteristics by which group-living animals signal danger as they flee. When one bird flies up, the rest of the flock usually follows. Many

FIG. 7.7. Some signaling movements originated in more elementary behavior patterns such as attacking, escaping, mating, and nest-building. Upright threat (*top*) consists of the assumption of an upright posture at the initiation of the attack, head held high for downward peck and wings partly opened for striking. If the opponent does not yield to the threat posture, the bird may attack, or, if the attack is inhibited by fear, the bird may turn sidewise in an appeasement posture, turning the beak upward.

Grass-pulling (*middle*) is often seen in territorial disputes. The bird pecks furiously at the grass as though attacking an opponent. This is a "redirected" attack. The movement terminates with a sideways flick of the head, such as is seen in normal nest-building. This element of the behavior is thus a displaced nest-building movement.

The kittiwake, a small gull, advertises the nest site to passing females by "choking," which consists of bending over followed by rhythmic up-and-down movements of the head (*bottom*). These are movements normally performed during incubation and infant feeding. They have here become ritualized as advertisements of territorial status. (After N. Tinbergen, *Scientific American,* July, 1962.)

ducks and geese, in which this characteristic is strongly developed and important in keeping the flock together, show conspicuously developed stripes or other markings on their wings. These become visible only when the animals take off. Such "specula," as they are called, constitute flashing signals that appear to release the flying-up behavior of their flight kumpans. A similar phenomenon in mammals is

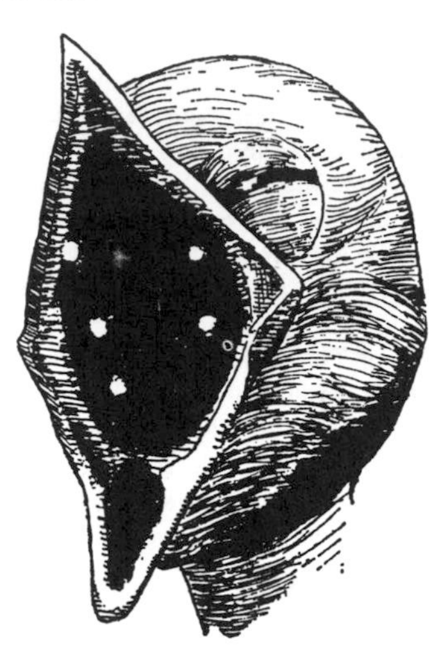

FIG. 7.8. The nestling of the bearded tit illustrates the remarkable development of conspicuous signals on the inside of the mouth which serve as releasers of parental feeding responses. There are four rows of pearly white conical projections set against a background of black. The interior is framed by a carnelian red border and lemon-yellow gape-wattles at the corners of the mouth. (Drawn from a photograph.)

the white rump and under-tail hair of ungulates, such as many deer and antelope. Our western pronghorn antelope shows this development to an extreme. The entire rump is covered with long white hairs. When the animal flees, these are stiffly erect, producing a bright patch that flashes in the sun as the animal dashes off (Fig. 1.1). This signal is said to be visible for long distances and to give warning to other members of the herd of the presence of a source of danger (Seton, 1953).

Of great interest for the general concept of social behavior is the way in which social releasers tend to evolve conspicuousness in behavioral movements by a process that has come to be called ritualization (Baerends, 1950; Hinde and Tinbergen, 1958; Blest, 1961). In this process the behavioral movements often take on a formal stiffness, slowness, or exaggeration that serves no function except to make them stand out sharply from ordinary functional movements. We have described previously the formal bowing, erection of head and neck, and bill-fencing that characterizes the courting and nest-relief ceremonies of gannets and other birds (Fig. 7.9). The "proud" postures with head held high and stiff walk of so many dominant animals, such as the master buck in the Indian antelope, may be viewed as ritualized threats (Fig. 7.10). Many of the intention movements mentioned earlier tend to evolve by ritualization (Daanje, 1950). Some threat postures, because of their exaggeration and slowness compared to normal fight reactions, serve as forms of communication. The stance, the carriage of the tail, and the facial expression of the wolf have been found to serve as signals of the status of the animal with regard

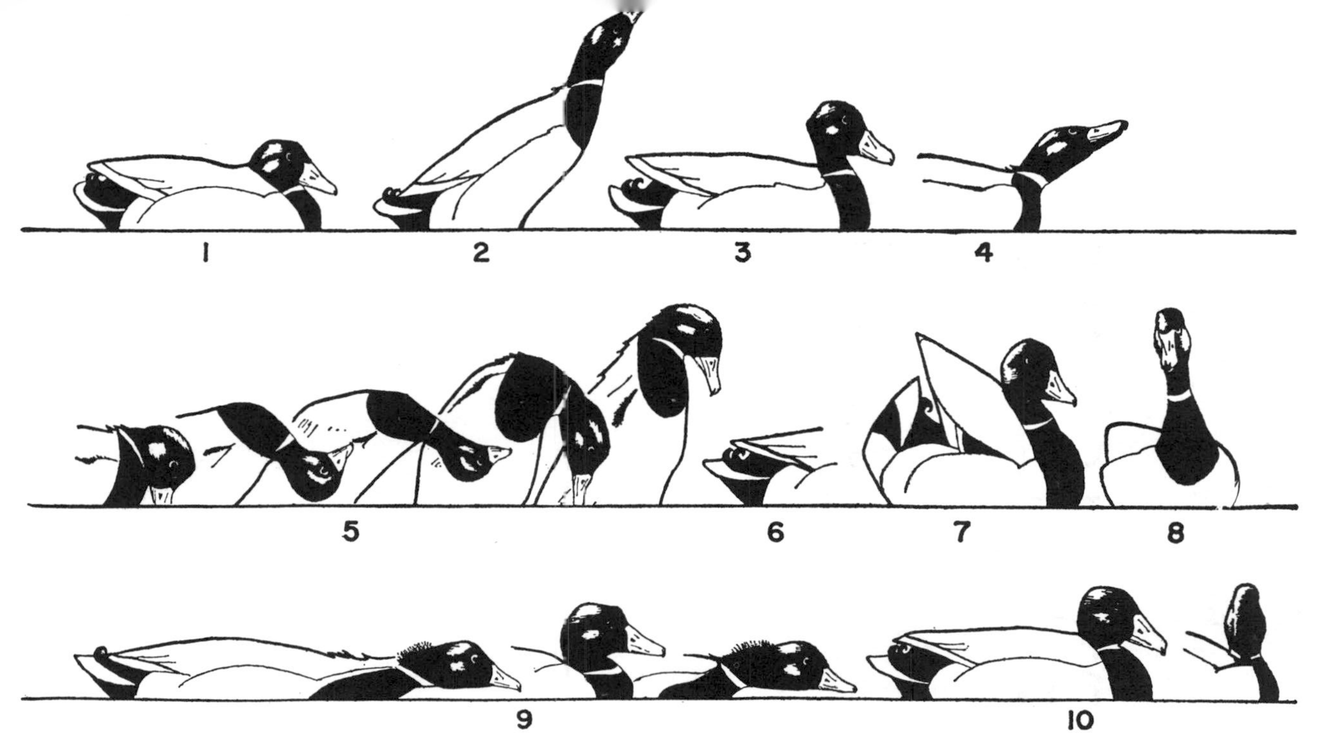


FIG. 7.9. If the courtship activities of related species are compared, many similarities appear which suggest their derivation from a common evolutionary origin. This is to say, behaviors may be homologized with each other just as structures are. This is especially clear in such a group as the surface-feeding ducks in which the elaborate and highly ritualized courtship patterns can be analyzed into a series of definite movements and poses which then are recognizable with modifications in different species. A sequence of courtship movements of the mallard is illustrated above. The descriptive terms applied to these actions by Konrad Lorenz are as follows: First the animal shows the "tail-shake" (*1*), followed by rising in the water to produce a "head-shake" (*2*). On falling back it shows a tail-shake again (*3*), followed by a "bill-shake" (*4*). An elaborate twisting of the head and rising out of the water called "head-up" (*5*) is followed by another tail-shake (*6*) then a "tail-up" (*7*), after which the male turns towards the female. (*8*). A "nod-swim" follows (*9*), then a turning of the back of the head (*10*). In other species, the individual components are so similar as to be readily identified and homologized with these. (Redrawn after K. Lorenz, *Scientific American*, 1958.)

FIG. 7.10. The master buck in the herd of Indian antelope leads the estrus female away from the others by pushing with the underside of his neck. When they get to the mating arena he courts her by a neck-stretching movement which appears to be a ritualized form of the herding movement by which he led her there. He does not actually touch her in this neck-stretching but repeatedly poses over her. The estrus female responds to the male's overtures by standing quietly, whereas the non-estrus animal moves away whenever approached by the master buck. Only after much courting does the male attempt mounting. (From W. Etkin, "Dominance Behavior in Black Buck" [motion picture].)

to dominance. These may be regarded as more or less ritualized autochthonous movements of dominance or submission. To some extent they are ritualized intention movements of attack or flight. (See Fig. 10.11.)

In discussing the concept of action-specific energy, we mentioned the idea that when an instinctual center accumulates large amounts of such energy, it tends to be released either by vacuum activity or by "sparking over" to some other, usually related, motor center. Such sparking over we called "displacement." One of the most interesting aspects of the contemporary study of instinct is the discovery of the ways in which displacement activities tend to evolve into conspicuous sign stimuli by ritualization. One of the best examples of this is the analysis of displacement digging in sticklebacks. Tinbergen found that when patrolling his territory borders, a male often turns vertically and makes displacement movements of pecking at the floor. This was seen when he was faced by a formidable opponent, usually the neighboring territory holder. When the experimenter exaggerated the conditions favorable for this display by crowding the males, the displacement activity became vigorous and frequent enough to reveal its true nature. The animals were not merely picking up bits of the substrate, as though feeding or randomly discharging, but they actually excavated nests just as in true nest building. Tinbergen's interpretation of this is that when the animal is faced with an opponent, its fighting behavior is released; but when the opponent is outside the territory or favored otherwise, the discharge of aggressive fighting is inhibited and tends to spark over to the nest-digging activity, which is another prominent element of the animal's territorial behavior. In the stickleback, this displaced digging activity has evolved into a sign stimulus which serves to inhibit the aggressive activity of the kumpan, with the result that the mutual display of displacement digging keeps the aggressive tendencies of the territorial males from becoming self-destructive by excess fighting (Tinbergen, 1951).

From our present point of view the significant finding is that the displacement activity has become a signal between kumpans. As such, its efficiency has been sharpened by natural selective action leading to ritualization. Such evolution has been in the direction of making the movements stand out by their conspicuousness. The threatening stickleback turns so that the opponent sees his side view; he erects his ventral spines, but, most conspicuously of all, he pecks furiously at the sand in contrast to the mild functional level of pecking in real nest building. This vigor is evident in many other examples of displacement activity. Herring gulls and greylag geese show grass-pulling displacement activity when fighting. This activity is expressed with much more "animus" than when similar action is taken in nest-building activity. The food-pecking activity that roosters often show as a displacement activity when fighting is similar.

Not only are such displacement activities made conspicuous by the extra vigor and by special postures, but it should be noted that the very inappropriateness of the action commands attention. To the examples mentioned above we can add the displacement activity of sleeping shown by many wading birds, such as avocets, in conflict situations.

In Tinbergen's view, many of the individual movements that constitute the courtship sequence of the stickleback are displacement activities. They are brought on by a conflict of drives in the male but serve as signals for the release of appropriate behaviors by the female at each point. Tinbergen regards the zigzag dance by which the male leads the female to the nest as an alternation of attack movements toward the female, who is violating his territory, and retreat movements toward the nest, determined by sexual releasing stimuli from the female. When he arrives at the nest, his posing with head partly in nest and later his poking of the female when she is partly in the nest are interpreted as displacement of the normal functional fanning movements by which the male aerates the eggs in the nest. Tinbergen regards these courtship movements as being derived from related activities of territory guarding and parental care which have been displaced and ritualized to form signals that release the appropriate behavior of the kumpan. The attractiveness of this hypothesis in providing a clear evolutionary explanation of these otherwise strange courtship behaviors is evident. It is discussed more critically in Chapter 8 (see Fig. 8.4).

It should be noted that derived movements, whether developed from intention movements or displaced activities, tend to take on the same regular and fixed character noted in instinctual activities in general. Thus each species tends to have fixed forms for the displaced activity (avocets sleep, geese pull grass, etc.) when fighting is inhibited. In other words, the process of sparking over is not random, happening now in one way, now in another. In part, we can see that this is the result of the fact that the outflow of energy in displacement is to another activity related to the first—as nest building is related to territory defense in the stickleback. Of course, such fixity of response facilitates the evolution of the response as a social releaser; for, to function efficiently as a stimulus in social communication, there must be a one-to-one correspondence between the displaced activity and the activity it is displacing.

This summary of the evolution of animal signals is based on the earlier theoretic writings of the ethologists. In reading the next chapter, it will be noticed that Tinbergen now gives greater prominence to intention and other derived movements than to displacement. Though recognizing the importance of displacement, Armstrong (1950), Hinde (1961), and Rowell (1961), as well as Tinbergen (Chap. 8), seek other explanations of the physiology of such behavior than that of overflow of action-specific energy as described above.

## Learning and Socialization

In the discussion of kumpan theory above we arrived at the concept that animals are in some cases held together in appropriate social interactions by releasers, either innate or imprinted, which evoke appropriate interactions between them. Such a concept can apply, of course, only to the highly stereotyped behaviors characteristic of instincts. Examples of such behavior are readily found in lower vertebrates and were extensively commented on above.

There are many other behaviors to which the kumpan idea cannot be applied. These include the less stereotyped social behaviors, particularly common in mammals but found also in lower vertebrates wherever individuality and specificity in social relations are common. For example, in Chapter 1 we pointed out that social organization based on social dominance with its correlates of hierarchy and closed-group formation depends upon recognition of individuals. Pair formation, discussed in Chapter 5, likewise is based upon the ability of the mates to respond differentially to each other as individuals. Similarly, territorial behavior requires recognition of specific areas. These are all characteristics which must be learned in the course of an individual's own lifetime. Early experiences in a group are often of great importance in socialization in animals (Figs. 7.11, 7.12, 7.13, and 7.14). This will be discussed in detail in Chapter 9. We may characterize such learning as familiarization with the environment, social and physical. Its importance in social behavior was sufficiently emphasized earlier in this book. Here we may ask how familiarization is attained and maintained.

We may note first that the capacity for each type of familiarization is genetically determined as a species characteristic. Thus each species has, as we pointed out above, its own innate propensities to learn in this respect. In birds with well-developed social dominance, individual recognition is quickly learned and long maintained, and similar relationships were seen to obtain in territoriality, mate recognition, and parent-offspring relations, when these phenomena were discussed earlier. Thus an essential factor in this type of socialization is the sensitivity of the animal to the particular learning experience.

Related to this sensitivity is another characteristic of the behavior of animals in which inter-individual familiarization is important. Animals commonly display behaviors which serve to reinforce learned relations. We have already emphasized this aspect in dominance behavior, by pointing to the patterns of dominance display behavior shown by the alpha animal in hierarchies (Chap. 1). We also interpreted mutual greeting or post-nuptial courtship ceremonies as tend-

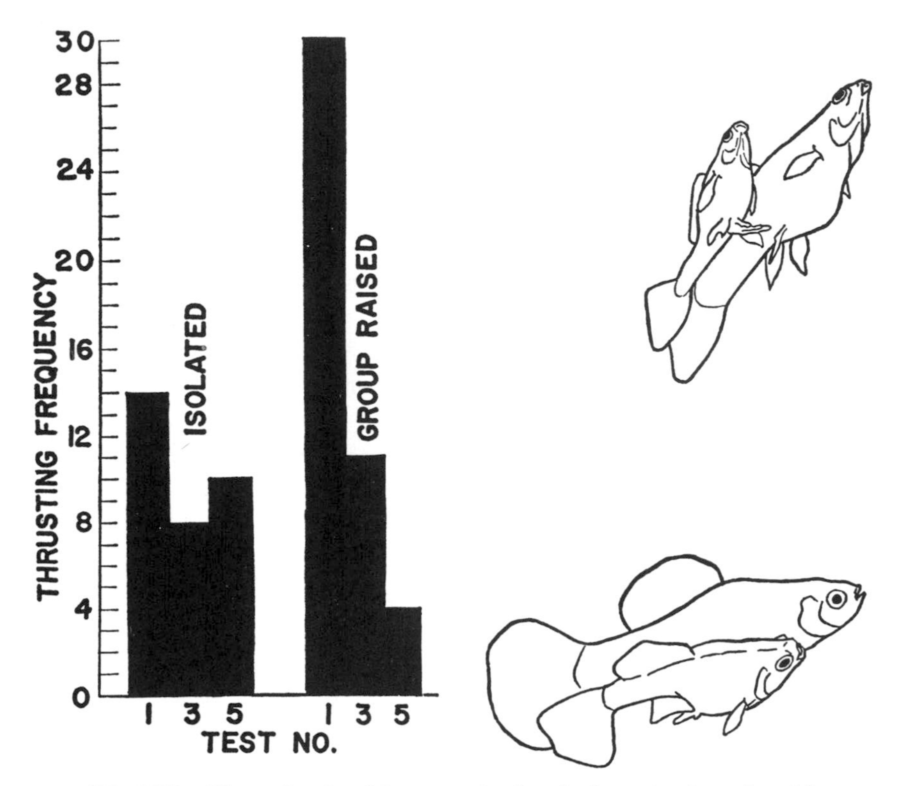


FIG. 7.11. The male platyfish courts the female by swimming alongside her and thrusting his gonopodium toward her genital aperture with the aid of the pelvic fin (*upper right*). Copulation is attained only occasionally in contacts which last several seconds (*lower right*). Males raised in isolation from species mates showed the essentials of courtship behavior but were partially inhibited in the expressions of some of their courtship activities in their first tests but not later. (After Shaw, *in* E. Bliss (ed.), *Roots of Behavior*, 1962.)

ing to maintain the learned relationship of mates (Chap. 4). In addition to these there are other forms of behavior that appear to function importantly in familiarization and which we have not yet discussed.

Some species of monkeys and other primates keep themselves quite free of parasites, debris, etc., by picking assiduously through their own and their companions' fur with their fingers and removing foreign matter. This grooming activity, especially between members of the group, seems far more extensively pursued than necessary for its toilet function. The animals are quite content to groom one another for considerable periods without receiving any tangible reward. Mutual grooming, grooming of dominants by subordinates, and offers to groom or be groomed as means of avoiding conflict situations are all

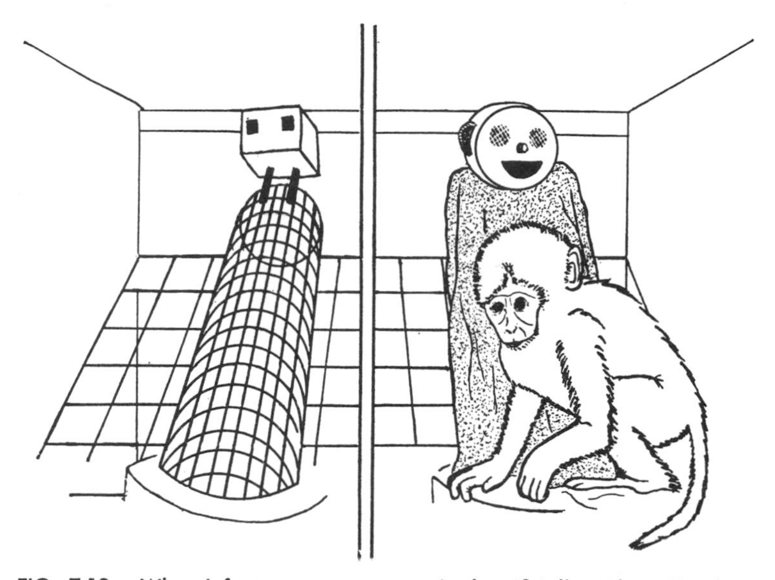

FIG. 7.12. When infant macaques are raised artificially without "mothering" of any kind their behavioral patterns are abnormal. In particular, they show disturbed, fearful behavior toward any new factor in the environment and are unable to adjust to companions. If a mother substitute is provided, this helps in permitting normal behavior development. The terry-cloth covered mother-surrogate (*right*) affords some comfort in fearful situations, but the wire mesh model (*left*) apparently does not. This is true whether the infant receives its milk bottle at the wire or cloth model. If raised with another infant, the animals cling together and thereby achieve considerable normalization of their behavior development. (After H. Harlow, *in* E. Bliss (ed.), *Roots of Behavior,* 1962.

commonplace among monkeys and apes. Some birds (for example, the yellow-eyed penguins) show a similar if less extensive mutual preening ceremony; but, with a few such exceptions, grooming is a distinctive behavior of primates (Nissen, 1951).

Such grooming resembles other types of persistent exchange of attentions that characterize many learned relations of familiarity in animals. The billing and cooing of doves and other birds, the persistent sniffing of one another by dogs and rodents, and the repeated

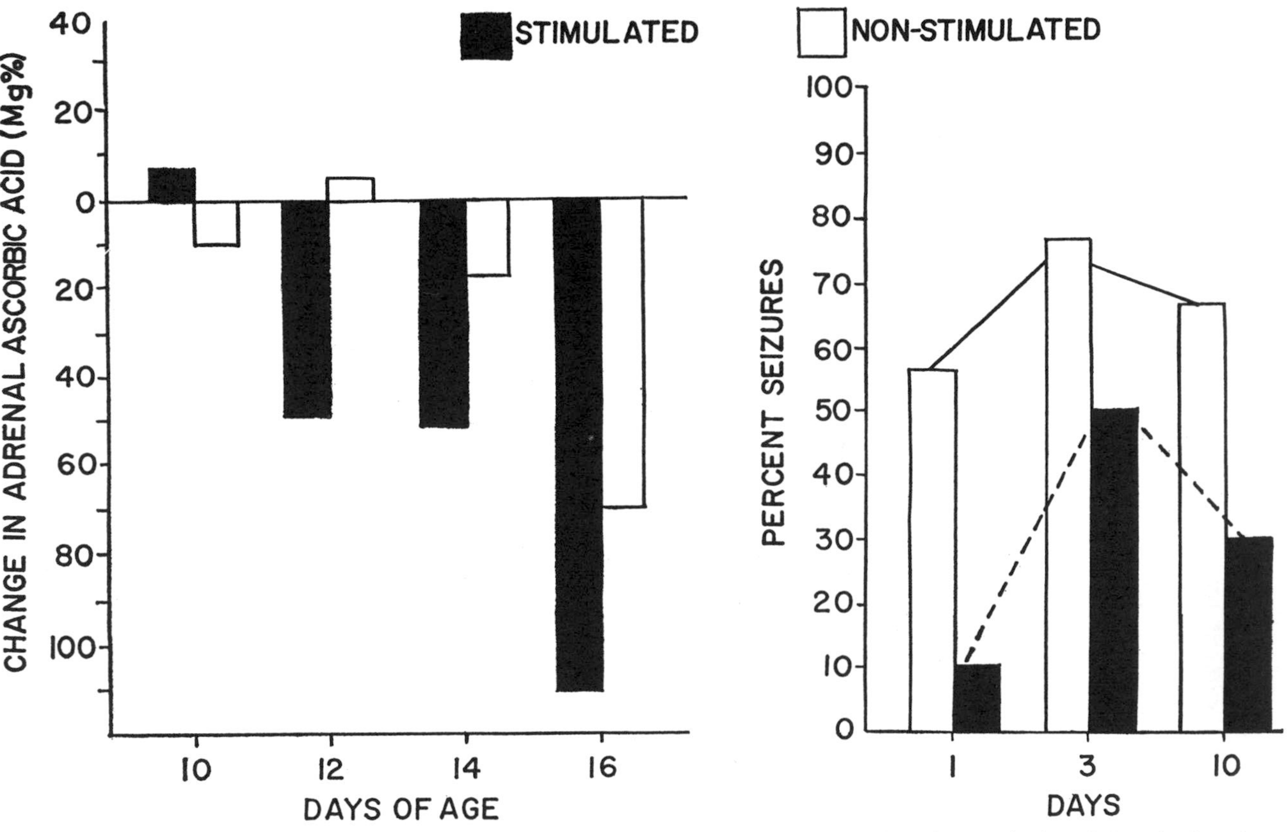


FIG. 7.13. The profound influence early experience can have upon animals is nowhere better displayed than in the effects that the simple procedure of handling the animal infant for a few minutes daily has upon the maturation of the stress response system of the adrenal. This may be measured by the amount of reduction in the ascorbic acid content of the gland when the animal is stressed by being put in the cold. As seen in the graph (*left*) the newborn rats show no significant response, but at twelve days of age the manipulated animals have a clearly developed response and continue to show a greater response than non-manipulated animals through the sixteenth day. Other measures of development such as time of opening the eyes and sensitvity to electrical seizures (*right*) are likewise affected. (After Levine, Alpert, and Lewis, *in* E. Bliss [ed.], *Roots of Behavior*, 1962.)

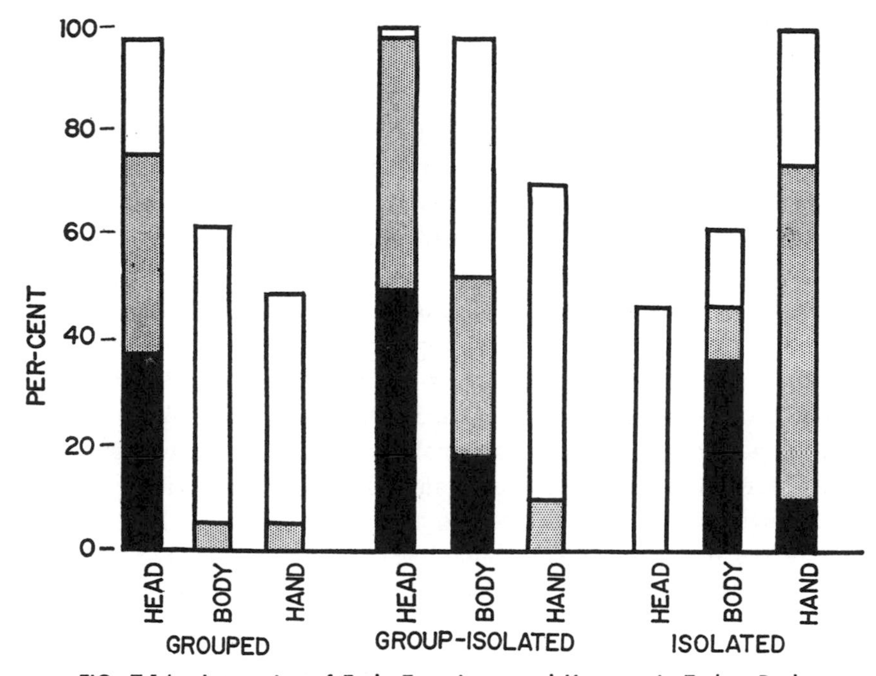


FIG. 7.14. Interaction of Early Experience and Hormone in Turkey Poults The interrelations of hormone levels and early social experience is well illustrated in an experiment on turkey poults. Some turkeys were raised in isolation, being hatched from eggs covered by aluminum containers to prevent visual or tactile experience during the hatching process in the incubator and kept in separate boxes thereafter. Three types of birds were studied at thirty-two days of age: (*1*) animals raised in isolation from other birds but exposed to the usual handling by the experimenter, (*2*) animals kept continuously in groups, and (*3*) animals raised in groups for five or ten days and then isolated. Since all were immature at the time of testing they were each injected with the same dose of male hormone to bring out their courting behavior. The test objects were a stuffed body of a poult, a poult head, or the human hand. The graph above shows the percentage of time during a five-minute test, spent by each type of experimental bird in making copulatory movements (*solid Black*), mounting attempts (*cross hatch*), or strutting display (*white*) toward each type of stimulus. It can be seen that the head was by far the most effective stimulus for the grouped animals, whereas the hand was best for the isolated birds. The importance of the nature of an animal's early experience for determining the nature of the stimuli releasing its behavior, is thus well illustrated. (After M. W. Schein and E. B. Hale, *Animal Behaviour,* 7 [1959].)

prodding of the females by male guppies all illustrate the tendency common in social animals to maintain persistent contact by reiterated exchange of mild stimuli. Many social animals keep in touch with one another by repeated sounds or visual signals. Such contact presumably helps to maintain the sense of familiarity so important to their social stability.

Social play is one of the most interesting of the ways in which animals appear to maintain familiarity with other group members. Though the young of passerines and penguins and certain other birds seem to play (Nice, 1943), the phenomenon is most conspicuous in social mammals. It is far more conspicuous in the young than in the adult. Young macaque monkeys seem to be as much taken up with play activity as human children. Lambs and other ungulate young play at gamelike activities resembling tag, king-of-the-hill, and other favorites of children. Play-hunting is common in carnivores. Play no doubt serves other functions (Beach, 1945), but it must help in establishing and maintaining learned social affinities. It is noteworthy that play is rare in the adult of most mammals where social relations are already well stabilized. However, in wolves, in which adult activities require the maintenance of a high level of co-operative behavior in hunting, play, especially play-fighting persists among adults. Play may thus be considered one of the techniques of reiterated stimulus exchange by which social animals maintain their familiarity with each other as individuals. It appears to be an important technique by which young social mammals learn their place in the group and develop appropriate in-group feeling. It is often conspicuous in maintaining pair relations in social mammals. In this respect it replaces the kumpan relation commonly serving these functions in lower vertebrates with more stereotyped behaviors.

## REFERENCES

ARMSTRONG, E. 1950. The nature and functions of displacement activities. *In: Physiological Mechanisms in Animal Behavior*. New York: Academic Press.

BAERENDS, G. P. 1950. Specializations in organs and movements with a releasing function. *In: Physiological Mechanisms in Animal Behavior*. New York: Academic Press.

BEACH, F. A. 1945. Current concepts of play in animals. *Amer. Nat.*, **79:** 523–41.

BLEST, A. D. 1961. The concept of ritualization. *In:* W. H. THORPE and O. L. ZANGWILL (eds.), *Current Problems in Animal Behavior*. Cambridge: Cambridge University Press.

COLLIAS, N. E. 1962. Social development in birds and mammals. *In:* E. L. BLISS (ed.), *Roots of Behavior*. New York: Harper & Bros.

CRAIG, W. 1918. Appetites and aversions as constituents of instincts. *Biol. Bull.*, **34:** 91–107.

DAANJE, A. 1950. On the locomotory movements of birds, and the intention movements derived from them. *Behaviour,* **3:** 48–98.

EMLEN, J. 1955. The study of behavior in birds. *In:* A. WOLFSON (ed.), *Recent Studies in Avian Biology.* Urbana, Ill.: University of Illinois Press.

GUITON, P. 1959. Socialization and imprinting in brown leghorn chicks. *Animal Behaviour,* **7:** 26–34.

HEDIGER, H. 1955. *Psychology of Animals in Zoos and Circuses.* New York: Criterion Books.

HESS, E. 1962. Imprinting and the "critical period" concept. *In:* E. L. BLISS (ed.), *Roots of Behavior.* New York: Harper & Bros.

HINDE, R. A. 1961. Behavior. *In:* A. J. MARSHALL (ed.), *Biology and Comparative Physiology of Birds,* Chap. 23. New York: Academic Press.

LACK, D. 1943. *The Life of the Robin.* Reprinted, 1953; London: Penguin Books.

LANYON, W. E. 1960. The ontogeny of vocalization in birds. *In:* W. E. LANYON and W. TAVOLGA (eds.), *Animal Sounds and Communication.* Washington, D.C.: American Institute of Biological Sciences.

LEHRMAN, D. S. 1953. A critique of Konrad Lorenz's theory of instinctive behavior. *Quart. Rev. Biol.,* **28:** 337–63.

LORENZ, K. Z. 1950. The comparative method in studying innate behaviour patterns. *In: Physiological Mechanisms in Animal Behavior.* New York: Academic Press.

———. 1957. Companionship in bird life and other essays. *In:* C. SCHILLER (ed.), *Instinctive Behavior.* New York: International Universities Press.

NICE, M. M. 1943. Studies in the life history of the song sparrow, II. *Trans. Linnaean Soc.* (N.Y.), **6:** 1–238.

NISSEN, H. 1951. Social behavior in primates. *In:* C. P. STONE (ed.), *Comparative Psychology.* 3d ed.; New York: Prentice-Hall.

ROWELL, C. 1961. Displacement grooming in the chaffinch. *Animal Behaviour,* **9:** 38–64.

SCHEIN, M. W., and HALE, E. B. 1959. The effect of early social experience on male sexual behaviour of androgen injected turkeys. *Animal Behaviour,* **7:** 189–200.

SCHNEIRLA, T. C. 1956. Interrelationship of the "innate" and the "acquired" in instinctive behavior. *In: L'Instinct dans de comportment des animaus et de l'homme,* pp. 383–452. Paris: Masson et Cie.

SETON, E. T. 1953. *Lives of Game Animals.* 3 vols. Newton Centre, Mass.: Charles T. Branford.

SMITH, S., and HOSKING, E. 1956. *Birds Fighting.* London: Faber & Faber.

THORPE, W. H. 1956. *Learning and Instinct in Animals.* Cambridge, Mass.: Harvard University Press.

TINBERGEN, N. 1951. *The Study of Instinct.* Oxford: Clarendon Press.

———. 1953. *Social Behaviour in Animals.* New York: John Wiley & Sons.

WHITMAN, C. O. 1919. *The Behavior of Pigeons,* ed. H. CARR. Washington, D.C.: Carnegie Institute.

YOUNG, W. C. 1961. Hormones and mating behavior. *In:* W. C. YOUNG (ed.), *Sex and Internal Secretions.* Baltimore: Williams & Wilkins Co.

NIKO TINBERGEN

# 8

# The Evolution of Signaling Devices

## Introduction

Co-operation between individual animals, by which some end is achieved which each of them alone could not attain (mating, spacing-out, raising of offspring, warning against predators, etc.) usually depends on interaction by means of a signaling system. One party—the actor—emits a signal, to which the other party—the reactor—responds in such a way that the welfare of the species is promoted. In his classic paper, "Der Kumpan in der Umwelt des Vogels" (1935)—which contains much more information than the much-abridged translations printed in the *Auk* (1937) and in Schiller (1957)—Lorenz organized and greatly extended our knowledge of such signaling systems; his paper has done much to stimulate research. Lorenz drew attention to the fact that many animal species possess a class of effector organs (brightly colored structures, mechanisms for movements and postures, scent-emitting organs, sound-producing devices, etc.) whose main, or often only, function is that of emitting signals. He pointed out, further, that such species also possess the specific responsiveness without which the signaling system would not work. Both these transmitting organs (which he called "Auslöser"—translated by American authors as "releasers"—but which, in order to avoid terminological confusion, will be called "signaling devices" in this chapter) and the receptory correlates are often characteristic of the species. Within the species they are often characteristic of sex and age classes (Figs. 8.1, 8.2). As a rule these signaling devices, even where they are movements or postures, develop normally even in animals raised in the ab-

Niko Tinbergen is university lecturer in animal behavior at Oxford University.

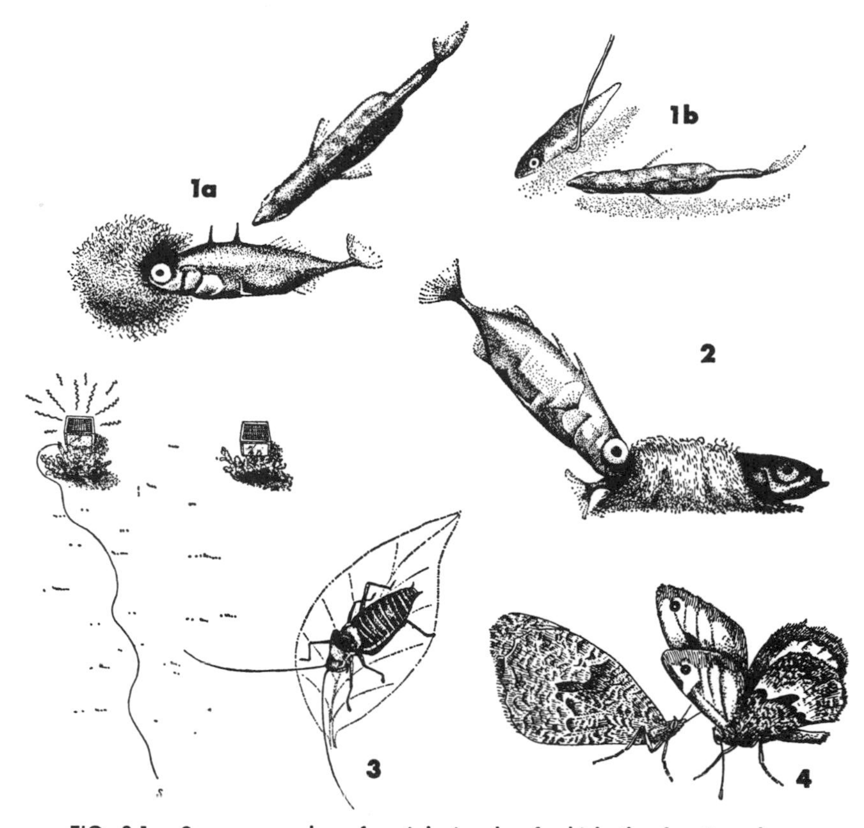


FIG. 8.1. Some examples of social signals of which the functions have been demonstrated in experiments: (*1*) Movement: (*a*) male three-spined stickleback showing nest entrance to female, and (*b*) the female responding to a dummy. (2) Touch: female three-spined stickleback spawning in response to the male's "quivering." (*3*) Sound: female ephippiger (*right*) and a track of a sexually active female (*left*) in response to the male's stridulation sounds. (*4*) Scent: male grayling performing the "bow," in which the female's antennae are brought in touch with the "scent fields" on the male's fore-wings. (After N. Tinbergen, 1953.)

sence of fellow species members (the "Kasper Hauser experiment" of German authors). They may also develop in animals raised with members of other species. The interesting phenomenon of some song birds' acquiring specific song patterns as a result of experience (Lanyon, 1957; Thorpe, 1958; Nicolai, 1959; but also Sauer, 1954) is rather an exception to this general rule. The specific responsiveness to signals, the development of which has naturally not been studied as extensively, may either be acquired under the influence of experience (Bergman, 1946; Immelmann, 1959) or it may develop independently of condi-

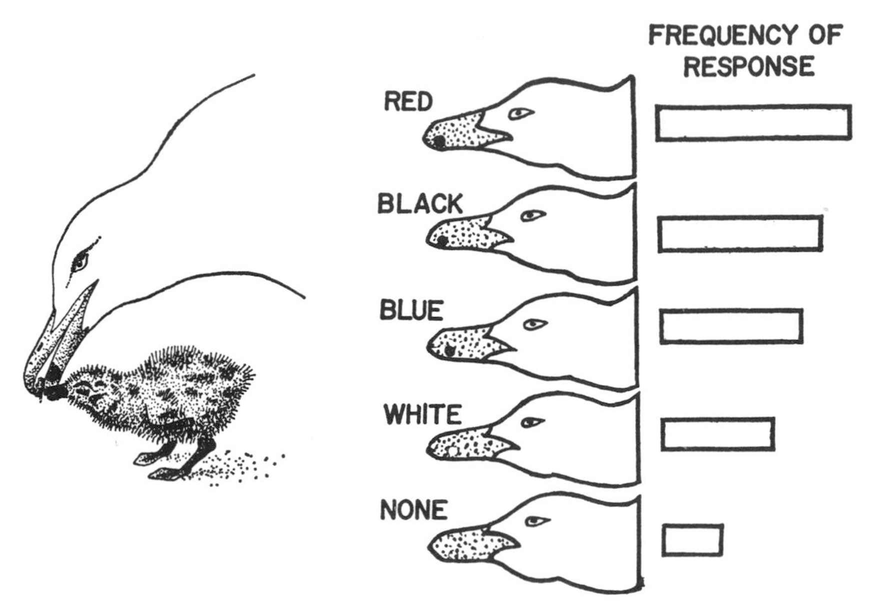


FIG. 8.2. Herring gull chick pecking at a flat cardboard dummy of an adult herring gull's head; this particular dummy has a black instead of a red color patch on the lower mandible (*left*). Experimental models were used in testing the function of the patch of color on the mandible of the adult herring gull in releasing begging responses in chicks. The length of the columns on the *right* indicate the relative frequency of chick response in each case. The evidence indicates that the red patch (the natural color) is the most effective signal. (After N. Tinbergen, 1949.)

tioning (Fabricius, 1951; Goethe, 1955; Dieterlein, 1959). Often a certain selectiveness develops in the response without experience and is then narrowed by learning. This seems to be the rule in imprinting and related phenomena (Fabricius, 1951; Tinbergen, 1953; Weidmann, 1958).

Such signaling systems are found, not only as parts of intraspecific co-operation, but also in symbiosis (Davenport, 1958) and as parts of defensive behavior against predators (Marler, 1955; Simmons, 1955; Blest, 1957). It is worth stressing that while the signaling function has been experimentally demonstrated in a number of cases, such evidence is still relatively rare and most of our knowledge derives from what could be called "semi-experimental observation" (for instance, systematic records of birds responding to calls given by individuals hidden from sight or of animals responding to quite inadequate objects which happen to have the same color as the adequate actor).

Much of this review will be concerned with signaling movements (as distinct from sounds, scents, bright colors, etc.), since it is these movements about which most is known from an evolutionary viewpoint. Unfortunately, reliable experimental data on movements is particularly scanty (mainly because they are not easily imitated in experiments), and there is, therefore, a great need for more experimental work. Nevertheless the systematic observational evidence on the signal value of movements and postures is substantial enough for our present purpose.

Before discussing evolutionary aspects, some examples of signals will be briefly reviewed. Some of these have been described in previous, more extensive reviews (Tinbergen, 1948; Marler, 1959). Some new evidence is added in this chapter, though completeness is not claimed.

## Some Selected Examples of Signals

### MOVEMENTS AND POSTURES

Movements and postures often convey information between individuals. Most of these seem to be perceived visually, though in the best-analyzed example—the "dance" of bees—this is uncertain. The dances of bees (for a summary see von Frisch, 1954) indicate to other bees the type, direction, and distance of a source of food. The type is indicated chemically; the direction is indicated by the straight part of the dance; and the distance, by the time spent "wagging" per circuit (von Frisch and Jander, 1957). The dance can also be used to indicate water, or even new living quarters (Lindauer, 1955).

Male three-spined sticklebacks show the nest entrance to a ripe female by turning on their side and prodding with the snout into the nest entrance; ripe females can be made to follow an appropriately moved dummy male and to attempt to "enter" at the place indicated even in the absence of a nest (Tinbergen, 1953).

Drees (1952) showed with the aid of dummies that the waving of the front legs which is characteristic of courting male Salticid spiders (*Epiblemum*) acts as a signal which suppresses the female's feeding responses. (See Fig. 4.9.)

In many other animals the effectiveness of movements as signals, though not proven by experiment, is strongly suggested by systematic observations. Thus female ducks have a special "inciting" or "sikking" motion of the head which makes their mates attack other males indicated by this motion (Lorenz, 1941, 1958). Kittiwakes can inhibit at-

FIG. 8.3. An example of a signaling movement, "facing away" in gulls, considered to be homologous throughout a group of related species: (*1*) in pair-formation sequence of lesser black-headed gull, (*2*) in a hostile encounter between two male common gulls, (*3*) in dispute over nest between kittiwakes, (*4*) in a juvenile kittiwake, and (*5*) in pair-formation sequence of black-headed gulls. (After N. Tinbergen, 1959.)

tacks by other kittiwakes by turning away the face (E. Cullen, 1957). See Figure 8.3.

One of the classic examples of a special posture acting as a signal is that reported by Allen (1934) who showed that male ruffed grouse copulate even with other males as soon as the latter adopt the prone position typical of a female willing to copulate.

### BRIGHTLY COLORED STRUCTURES

Brightly colored structures are often effective as signals. Sometimes the color is permanently on display, such as in the cere of the shell parakeet which is brown in females and blue in males. As early as 1926 Cinat-Tomson showed convincingly that when a male individual meets a strange bird, a blue cere elicits attack and a brown cere, courting. Once individuals know each other individually, this signal is no longer decisive; the responses become conditioned to a much more complex stimulus situation. Immelmann (1959) has experimentally demonstrated the signal value of colors in the zebra finch. The characid fish *Pristella riddlei* has a conspicuous black spot on its dorsal fin; these fish school much more readily with normal fish than with fish whose dorsal fins are amputated (Keenleyside, 1955). Blest (1957) showed, with the aid of models and with intact and operated animals, that the "eye spots" displayed by many usually camouflaged insects actually ward off song birds. Crane (1955) and Magnus (1958) demonstrated the signal value of color in the nymphalid butterfly *Argynnis paphia* and in heliconid butterflies.

### SCENT SIGNALS

Scent signals are undoubtedly widespread. A well-known example is the attraction of male moths by virgin females, which Butenandt (1959) reviews; Butenandt and his collaborators have succeeded in isolating the scent-producing substance in the silk worm moth. There can be little doubt that scent signals play an important part in territorial advertising of many mammals (Hediger, 1949; Fiedler, 1957).

### SOUNDS

Sounds have been shown in many cases to act as signals. Blair (1958) demonstrated that the croaking of various anurans attracts females. Perdeck (1958) reported the same finding for the "song" of the male *Chorthippus*, a grasshopper. Frings *et al.* (1955) showed that the "food call" of herring gulls attracts fellow members of the same species, and alarm distress calls of social birds have been shown to elicit specific responses in members of the flock (Busnel and Giban, 1960).

### TACTILE SIGNALS

An example of a tactile signal is the rhythmic-touch stimulus given by a "prodding" male three-spined stickleback to a female; ripe females can be made to spawn by imitations of this signal with a glass rod (ter Pelkwijk and Tinbergen, 1937).

The available evidence shows that signals are widespread; that they convey their information through a wide variety of sensory modalities; that they may be either intra- or interspecific; and that they may be of various types: simple or complex, immediate or delayed, brief or sustained. Recently there has been evidence which shows that signaling behavior can also affect somatic growth processes; thus Lehrman *et al.* (1961) have shown that the presence of the mate (and of nesting material) speeds up gonadotropin secretion in the female ring dove (see Chapter 6).

## The Evolution of Signaling Devices

### INTRODUCTION

It is important to realize that of the two main sources of information about the course which phylogeny must have taken, viz., fossil evidence and comparison of present-day species, it is the latter method only which is available to the student of animal behavior. This method is indirect; since we are concerned with non-repeatable events, no crucial experiments are possible. We have to rely more than workers in experimental sciences do on "probabilistic" reasoning. The successes of comparative anatomy and taxonomy—in the modern sense of "The New Systematics"—bear witness to the potentials of this method. Particularly, it allows us to trace (*a*) the origins of species and species-specific organs and (*b*) their further evolution in the course of improvement and taxonomic differentiation.

With respect to signaling devices, two valuable concepts have been worked out: "derived movement," and "ritualization." These concepts were adumbrated in the works of Selous (1901, 1933) and Huxley (1923) and worked out later by several authors. They facilitate the analysis of what are, to all intents and purposes, special cases of the general phenomenon of change of function (i.e., improvement and differentiation) during evolution (see Mayr, 1960). The method can be elucidated by some examples taken from morphology.

It is generally accepted that the wings of bats were originally forelimbs used for terrestrial ambulation like the forelimbs of other tetrapods but that they have been secondarily adapted to flying. This argument, as far as it is derived from comparative evidence, is based on the bat's resemblance to other mammals in a large number of other characters—from which it is concluded that they are derived from the same ancestors as other mammals. The fact that bats form a relatively small aberrant group suggests that they, and not the rest of the mammals, are the derived, secondarily changed group (a conclusion which is confirmed by paleontological and embryological evidence). Their general structure and topography indicate that the bat's wings are the homologues of the forelimbs of the other mammals, and this allows us to say that their evolutionary origin *is* the primitive mammalian anterior walking leg, from which they have diverged in the course of evolution. A study of their present function confirms that this divergence was adaptive, i.e., has made them suited to their new function.

Similarly, the great claws of the lobster can be recognized on comparative evidence alone to be the homologues of the first pereiopods of other crustaceans which have been secondarily adapted as food-grasping and food-crushing organs. In this case the recently changed organ exists side by side with the other pereiopods which have retained the structure of walking legs.

The same steps are taken in the reconstruction of the evolutionary history of many signaling devices. Many male ducks preen their wings as part of their courtship (Lorenz, 1941, 1958). The common descent of the dabbling ducks, based by taxonomists on morphological similarity, is strikingly confirmed when all known species-specific behavior characters are considered along with the structural characters (Lorenz, 1941). The occurrence, in all species of ducks, of preening—both "ordinary" preening functional in the care of the plumage and courtship preening—is accepted as proof of homology in the sense of derivation from preening of a common ancestor. In certain species courtship preening is very similar to ordinary preening; in some (such as the mandarin and the garganey) it is very different; others take up an intermediate position in this respect. In those species where the differences are great, courtship preening has all the characteristics of a releaser: the movement is striking and is made more conspicuous by the exposure of brightly colored structures, which are further enhanced by the male's turning the exposed structure toward the female. The conclusion is that courtship preening has been "derived" from ordinary preening as the lobster's claw has been "derived" from the first pereiopod, that it has acquired signal value, and that it has been secondarily adapted to this new function—has been "ritualized."

The rationale underlying the evolutionary interpretation of other signaling devices is similar. Unfortunately, the chain of evidence is in no case complete. In the following review the gaps in the evidence will be stressed along with the positive knowledge, which, while fragmentary, is extremely suggestive.

### THE ORIGINS OF SIGNALING MOVEMENTS

As Selous (1901, 1933) and Huxley (1923) have pointed out, the form of signaling movements and postures often gives a clue as to their origin. Thus many signaling movements resemble incomplete versions of movements which themselves have another function. For instance, many threat postures involve the first stages of fighting in which the weapons are brought into a position of readiness; birds point the bill at an opponent or lift the carpal joints; fish may open their mouths; many mammals bare their teeth. Selous and Huxley also recognized that other, seemingly irrelevant movements might have provided the "raw material" from which signals have been developed. Modern studies of the motivation underlying signal movements have suggested how such movements have "become available" as raw material, and it is becoming increasingly clear that many, if not all, signal movements are derived from expressions of motivational conflicts. We shall therefore first consider questions of motivation.

*The ambivalent motivation underlying signal movements.*—The question "What makes an animal perform a signaling movement?" has been studied most intensely in displays of birds, particularly in agonistic displays, in courtship "antics," and in distraction displays. Three methods have been applied. (For a discussion of these methods and for other references see Tinbergen, 1960).

First, the *motor pattern* or *form* of display itself was analyzed. Male sticklebacks which occupy an individual territory in spring intimidate intruding neighbors by swimming toward them, after which they retreat; these movements are performed in quick alternation. The forward motion can be recognized as an incipient attack; the retreat is the opposite, withdrawal from the rival. The upright-threat postures of a male gull are characterized by (*a*) stretching of the neck and downward pointing of the bill, and (*b*) lifting of the carpal joints. Both are elements of the movements of attack. An attacking gull either pecks down at its opponent or beats him with the folded wings. At the same time such a gull, which usually starts off by advancing toward the rival, stops before attacking him and often turns its side toward him or even turns away altogether. This is correlated with sleeking part of the plumage. In both cases we can recognize elements of alarm and of escape. Thus this method is based on recognition of the form of the movement or posture and of the similarity of its components to components of other known behavior patterns.

Second, postures alternate in quick succession, even in constant conditions, with certain other behavior patterns and not with others. Thus agonistic postures, such as the two mentioned, are found to alternate with full attack (biting in the male stickleback, either pecking or wing-beating or both in the male gull) and with full escape but not with other behavior patterns. This is taken as independent evidence for the conclusion that the posture is the consequence of the simultaneous arousal of attack and escape.

The third method is based on an analysis of the environmental situation in which the postures occur. This is best known in agonistic encounters. In the case of two neighboring territory holders—males A and B—holding territory a and b, respectively, A elicits attack from B when both are on territory b, but it elicits escape when both are on territory a. The opposite is true of male A: B elicits attack from A when both are on territory a and escape when both are on territory b. Agonistic postures which intimidate or contain the neighbor are always shown in a zone between territory a and b, where both attack and escape are aroused, but neither dominates the other.

The conclusion based on evidence obtained by these three methods shows that agonistic signals arise as a consequence of the simultaneous arousal of two major behavior systems: attack and escape. The animal is said to be in a motivational conflict, or an ambivalent state.

In other cases the underlying motivational conflict is of a different kind. Some postures, such as facing-away from another bird (shown by all gulls), arise from the simultaneous arousal of escape and any tendency to stay—this tendency can, but need not be attack (Cullen, 1957; Tinbergen, 1960). Tail-flicking in many passerine birds signifies a conflict between flying and hopping (Andrew, 1956). Many courtship activities of territorial species are consequences at least in part of the simultaneous arousal of three tendencies: (*a*) approaching the mate (either for copulation or, in species with a "bethrothal" period, for just being together), (*b*) attacking the mate, and (*c*) fleeing from the mate (Tinbergen, 1954; Hinde, 1955; Morris, 1956). For our present purpose it is not necessary to examine in detail how the conflicts are elicited, or exactly how they cause the movements which we observe; the significant fact is that many signal movements are nearly identical with those movements which appear in animals as the consequence of dual or multiple motivation. A discussion of the categories of such movements that have been distinguished so far and that have apparently served as the raw material for ritualized signals follows.

*Successive combination of heterogeneous components.*—Two signaling movements of the three-spined sticklebacks offer clear examples of the successive combination of heterogeneous components. The alternation between incipient attack and incipient escape movements discussed above is one. This type of alternation is extremely common in agonistic encounters in many animals. A second example is the stickleback's "zigzag dance" (Fig. 8.4) by which it stimulates the female to approach (Pelkwijk and Tinbergen, 1937). The dance consists of intermittent elements—the "zigzags"—each of which consists of a movement away from the female followed immediately by one toward her. The movement away—the "zig"—often continues into that of "leading," a special type of rapid, smooth swimming toward the nest, which stimulates the female to follow; the other part—the "zag"—often develops into an actual bite. This suggests that the zigzag dance originated as a succession of sexual and aggressive behavior components. However much the movement seems ritualized, the occasional "lapses" into immediate leading or immediate attack indicate that the motivation contains elements of sexual and aggressive factors.

*Simultaneous combination of heterogeneous components.*—A good example of the simultaneous combination of heterogeneous components is the posture which black-headed gulls adopt at the end of the "meeting ceremony." Each bird stands in the upright posture, which contains an element of attack (raising of the carpal joints, which is the preparation for wing-beating shown only in aggressive birds) and elements of escape (lateral orientation and facing-away, which are part of the escape pattern).

*Compromise movements.*—Andrew (1956) has called attention to the fact that a movement caused by ambivalent motivation may often be intermediate between the two conflicting movements. The clearest example is the phenomenon of turning away from another animal in conflict situations (Fig. 8.3). If the turning-away movement is compromised in agonistic encounters, there may be a gliding scale from facing to standing entirely turned away (Tinbergen, 1959) or fixed assumption of the sideways stance as in many fishes (Lorenz, 1935). Such compromise movements also occur in courtship, for instance in weaver finches (Morris, 1956) and in black-headed gulls (Tinbergen, 1959).

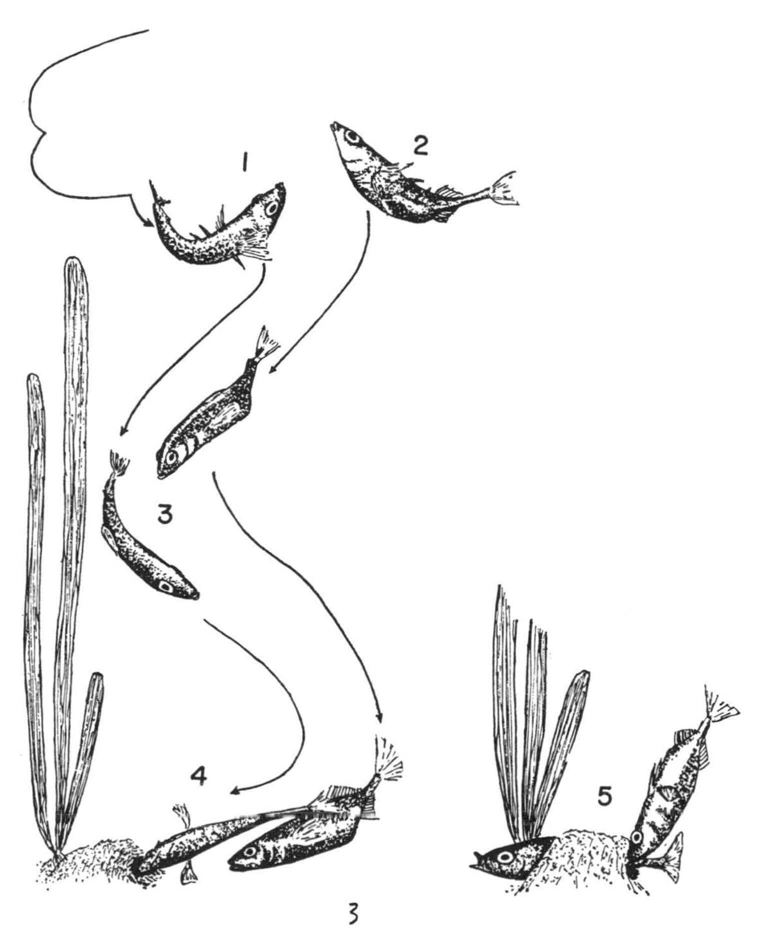


FIG. 8.4. The mating behavior of male and female sticklebacks was analyzed in terms of social signals. The appearance of the female evokes aggressive approach of territorial male (*1*). By turning upward, revealing her silver belly swollen with eggs (*2*), the female inhibits the attack of the male and releases his turning toward the nest (*3*). She follows him to the nest into which he pokes his head (*4*) and then withdraws. She enters the nest and stops. He then taps her tail with a trembling motion (*5*). This releases spawning of the eggs and swimming out of the nest by the female. The male then enters and fertilizes the eggs. Experimental analysis reveals this chain of behavioral responses to be largely dependent upon a succession of social signals. Each step of the sequence serves as the stimulus to release the next movement. (After N. Tinbergen, *The Study of Instinct*, 1951.)

*Redirected movements.*—As Bastock, Morris, and Moynihan (1953) pointed out, an animal aroused to attack another animal but inhibited from doing so may perform the full attack behavior but aim it at another individual or even at some less appropriate object. This is certainly common in gulls, but many other animals, such as dogs and monkeys manifest such redirected movements. Manley (1960) has convincingly argued that the "swoop and soar" of a male black-headed gull is a ritualized form of a redirected attack (Fig. 8.5). There is further evidence to support the conclusion that even the aerial dis-

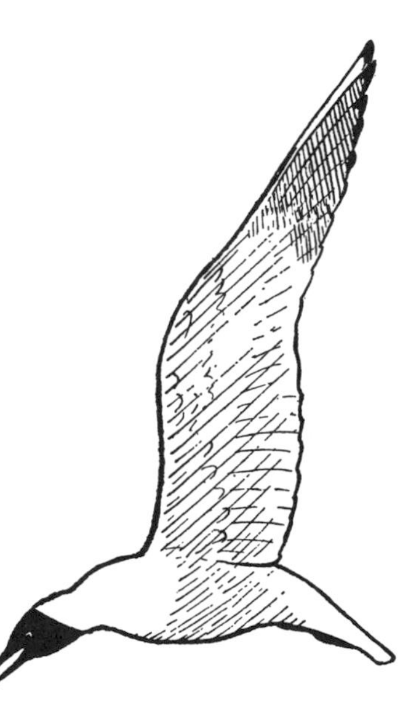

FIG. 8.5. A signaling movement considered to be derived from a "redirected attack": the "swoop-and-soar" of the black-headed gull. (After Moynihan [1955] from N. Tinbergen, 1959.)

plays of terns may be extremely ritualized derivatives of a redirected attack (J. M. Cullen, 1960).

*"Displacement" or "extraneous" movements.*—Movements similar to other movements shown in different functional and at least partly different motivational contexts often appear in conflict situations (Tinbergen, 1959; Rowell, 1961). As in the other categories, little is known about the exact causation of such displacement activities; certainly my original hypothesis (Tinbergen, 1940, 1959)—that they are due to a "surplus of motivation"—an idea also implicit in Makkink's original term of "sparking over" (Makkink, 1936)—was ill founded. A more likely explanation is based on the idea that two simultaneously aroused behavior systems, while mutually inhibiting each other, also inhibit each other's inhibitory effects on other behavior systems and thus pro-

duce displacement activities by disinhibition. This agrees with the fact that such displacement activities are always enhanced by the same factors which produce them as "ordinary" activities (van Iersel and Bol, 1958; Rowell, 1961). The occurrence of such movements in conflict or in general thwarting situations, however caused, is a fact (even though some movements previously described as displacement activities are no longer considered deserving of the name). There is also little doubt that many of them, such as the courtship preening of ducks and pigeons, serve as signals and have as such been ritualized.

Morris (1955) has shown that the movements of feathers (sleeking, fluffing, and ruffling), which play a part in so many displays of birds, are often highly similar to feather movements in the service of heat-regulation, which are under autonomic control. He also stressed the fact that many other autonomic responses (alimentary and respiratory) tend to appear in thwarting situations. Diagnostically they fall into the category of displacement activities. The cause of their appearance in these situations is again unknown, and it cannot be overemphasized that the terms "displacement activities," "irrelevant activities," and "extraneous activities" indicate unsolved problems, rather than physiological explanations.

Whatever their causation, many of these effects seem, as Spurway and Haldane pointed out in 1953, to have provided "raw material" for signals. Morris suggests some such origin for the flushing of bare-skin patches, as in the turkey; the use of defecation and urination in the service of territory-marking by mammals; inflation displays such as that of the frigate bird, the sage hen and related species, and bustards; scent-signaling in mammals which may have originated as sweating; and feather- and hair-raising, often of particular areas, which are elements of many signaling movements.

The classification given above is descriptive and provisional—descriptive in the sense that it does not intend to suggest explanations of the mechanisms involved (a classification according to mechanisms once these were known might well cut across the present one without "invalidating" it), provisional because it cannot be claimed that other types of raw material might not be discovered in the future. It should also be pointed out that many signaling movements seem to be a combination of elements belonging to different categories. Thus feather-raising may be combined with general sideways orientation, as in jays (Goodwin, 1952) and pheasants (Schenkel, 1956, 1958).

The important point here is that many signaling movements seem to be derived from thwarting and particularly motivational conflicts, the various behavioral consequences which have in some way acquired signal value and have, through being exposed to new selection pressures, been secondarily specialized as increasingly effective signals.

### THE RITUALIZATION OF SIGNALING DEVICES

It has already been said that a comparative study of signals within a group of closely related species reveals that many of them are similar to one another and to what is inferred to be their common origin. However, they are by no means *identical.* The differences are often small, but equally often they are considerable. They may, in fact, be so pronounced that if one confines one's studies to a single isolated species, the origin of a signal may be totally obscure. The fact that such signals are often linked through intermediates with related species' signals, the origin of which seems clear, is considered justification for the conclusion that they must all be homologous, i.e., derived from the same movement in the common ancestor. This means that any aberrant signal must have diverged from the less aberrant ones and from the common origin of both. A study of the functions of these apparently aberrant movements, i.e., their effectiveness as signals, and of the requirements of the receptory correlates of the reactor (strong stimulation of the sense organs involved) has led to the conclusion that this evolutionary divergence away from the origin was adaptive and has made these movements increasingly suited as signals. It is this inferred evolutionary process of increased adaptation to the signaling function which is called ritualization; as such it is a special case of adaptation to a newly acquired function.

Several authors have classified, on the basis of our meager knowledge, the various known ways in which signal movements appear to have been changed during ritualization (see, e.g., Morris, 1957; Tinbergen, 1960; Blest, 1961). Many of the same strictures apply to some of these classifications as apply to the explanations of the origins of derived movements: apart from the fact that they cannot possibly be claimed to be exhaustive, the criteria of classification cannot—in the present state of inadequate knowledge—refer to the neurophysiological mechanisms of these behavior patterns (let alone to their alteration in evolution). The classification given here is not intended to be complete but merely lists some of the more striking types. This list is based partly on formal analysis of the motor patterns observed and partly on what little is known about their dependence on external stimuli and on motivation. Its aim is to point out the variety of phenomena to be investigated.

The extent, or amplitude, of a movement has often been increased or decreased in the process of ritualization. Increase, often of components of a total movement, leads to exaggeration. For instance, tail-wagging movements in fish, upward movements of the head in ducks and gulls, wing- and tail-spreading and opening of the bill in many birds have often reached grotesque proportions. Often other components of movements have been reduced in scale and often completely dropped out so that the resulting ritualized signal is a much simpler movement than was the original. Sounds, particularly those of alarm and those functioning at great distances, have become extremely loud. Very often a movement has become rhythmic, i.e., what was originally a single or slightly repetitive movement has developed into a long series; the "choking" movements of gulls, their "head-tossing" movements, and the wing-flapping of cormorants are other examples. Repetition also plays a part in many sounds.

The speed of movements has often changed in the process of their ritualization; they have sometimes been speeded up (often combined with "freezing" in a striking posture at the end, such as in facing-away in black-headed gulls and in male ducks) and sometimes slowed down to an "emphatic" movement. Often parts of a movement are speeded up, other parts slowed down, and either component can in addition have an increased amplitude. A beautiful example of such complex effects is provided by the fiddler crabs (Crane, 1957). The main advertising movement of the males is a waving movement of the greatly enlarged claw; the type and speed of this movement is strikingly different in different species (Fig. 8.6). These crabs also illustrate another aspect of ritualization: the structures moved have become greatly enlarged and often brightly colored. Such bright coloration of displayed parts has gone to quite amazing extremes in many visually displaying animals—birds of paradise, pheasants, ducks, and many fish. The sound-producing organs of stridulating insects and of song birds are likewise examples of structural adaptation to ritualization.

Often ritualization results in combination (in the form either of a simultaneous mosaic or of a fixed sequence) of components. While the elements may be typical of a genus or an entire family, each species may have combined them into different compound displays. Beautiful examples of this phenomenon in ducks have been reported by Lorenz

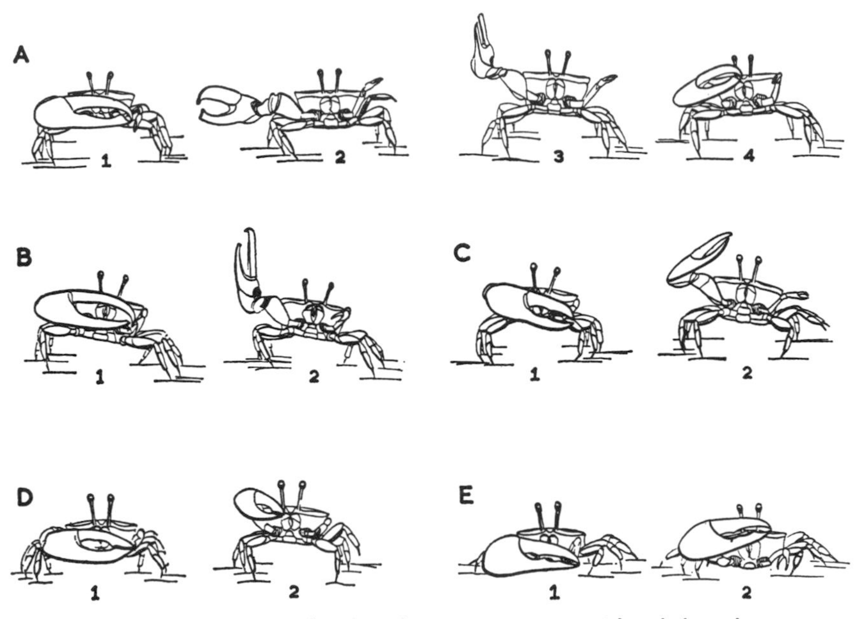


FIG. 8.6. Divergent radiation of a movement considered homologous throughout a group of related species: "claw-waving" in male fiddler crabs (Genus *Uca*).

A, 1–4: *Uca lactea* (Fiji)
B, 1–2, *Uca pugnax rapax* (Venezuela)
C, 1–2: *Uca zamboangana* (Philippine Islands)
D, 1–2: *Uca signata* (Philippine Islands)
E, 1–2: *Uca rhizophorae* (Singapore)

(After J. Crane, 1957.)

(1958); fixed sequences and simultaneously combined components are also found in the courtship of gulls.

Displays are as a rule selective responses to individuals of a certain class (such as the mate or the rival). This of course means that the mechanisms for this selective responsiveness must have changed along with the evolution of each species, at least in those cases where learning does not account for species recognition.

Morris (1957) has called attention to the fact that many signal movements have developed a "typical form"; many displays appear in roughly the same form under a wide variety of motivational situations which may originally have evoked a gliding scale of movements. Morris illustrates this with the ruffled-feather courtship posture of a male cutthroat finch, which has developed this all-or-none character. In many movements arising from a motivational conflict (for instance, the attack-escape conflict) only one posture out of a gliding scale of many relatively more aggressive and relatively more timid movements is assumed in a wide range of motivational situations.

These few examples show that ritualization has in a variety of ways resulted in making a signal more conspicuous (more strongly stimulating) and in making it stable (having a typical form). This stabilization seems in many cases to be closely connected with intersignal distinctness; in other cases signals of different species have been stabilized convergently. These aspects will be discussed in the next section.

### THE SELECTION PRESSURES INVOLVED IN RITUALIZATION

We have seen that comparative description of signals together with a study of their motivation have given us an idea of their evolutionary origin and that interspecific comparison combined with a study of signal functions leads to a description of their alleged ritualization. Assuming that our reasoning has been correct, can we guess what selection pressures must have been responsible for ritualization? The paucity of experimental work on survival value renders such guesses doubly hazardous, yet some hypotheses deserve mention.

*General improvement.*—It is obvious that much of the efficiency of signals is due to their power to provide strong stimuli. It cannot be accidental that many properties of signals (particularly those which are visual and auditory) fit so well the requirements of the sense organs by which they are to be perceived. Suddenness of movement, enlargement such as attained by inflation and feather- or hair-raising, rhythmic repetition, exaggeration, the development of brightly colored organs with striking and sharp contrast lines and of sudden, loud or staccato calls—features exactly opposite in many respects to those of visual and auditory camouflage—all fit these requirements. The conclusion that these properties are the consequence of direct selection for increased effectiveness is certainly not far-fetched.

It seems quite probable at the same time that in many cases the limit of this improvement has been reached. This can only be understood if it is realized that these limits are determined by opposing selection pressures. Many small animals have predators whose task is made easier by the very stimuli which signals are adapted to provide, and each species has, therefore, limits beyond which it cannot go, a fact already stressed by Huxley (1923). Looking at the problem in another way, the fact that so many signals have been allowed to become extremely elaborate and conspicuous is convincing proof of their importance. Their elaboration and conspicuousness have been made possible by such safeguards as the ability to change, e.g., color, through the mobility of chromatophores, or shape, through folding and spreading of wings, fins, etc., or through inflation and deflation of pouches. Periodic molts, etc., offer long-term solutions to this conflict. Marler (1959) has pointed out that the alarm calls of many song birds have been partly molded by the need to avoid giving certain predators directional clues.

Although predator pressure is certainly a limiting factor of great importance, the requirements of other functional systems such as feeding, respiration, or locomotion in general may also have set limits to the development of signals.

The evolution of signals has not been governed merely by selection for conspicuousness and counter-selection setting limits to improvement, however. Signals, like all adaptive features, show numerous functional interrelationships with other parts of an animal's equipment. One could classify these relationships roughly as follows:

*Intraspecific distinctness.*—There is ample evidence to show that the signals occurring within a species have mutually influenced one another so as to eliminate ambiguity. The most striking example of this is the distinctness, in territorial species in particular, of agonistic (threat, or distance-increasing) and appeasement (or distance-reducing) signals (Tinbergen, 1959). Thus in gulls various postures which involve pointing the bill (the main weapon) at the opponent are used as threat, whereas aggression is suppressed and co-operation made possible by facing-away. The striking intersignal distinctness which is the rule in other visual and auditory signal systems may well have been affected by a similar selection aimed at reduction of ambiguity.

*Interspecies distinctness.*—This seems to have been promoted particularly in those signals which bring the sexes together for mating. Blair's work (1958) has demonstrated this convincingly for various Anura, and Perdeck (1958) has beautifully demonstrated how sexual isolation between two sibling species of grasshoppers (*Chorthippus brunneus* and *C. biguttulus*) is effected by the distinctness of the songs of the males. In the Anura there is evidence showing that this divergence has been promoted by sympatry (occupation of the same area by two species). This phenomenon could be compared to the direct effects of competition in other functional spheres, such as feeding, of which the famous "Darwin's finches" of the Galapagos offer a striking example (Lack, 1947).

*Interspecific similarity.*—The original notion of the "arbitrariness" of signals as emphasized by Lorenz (1935) implied that signals were particularly valuable as taxonomic characters indicating affinity. Marler has shown (1957) that this early concept must be revised to a certain extent, since convergence does occur in signals as well as in all other characters. Thus the hawk-alarm calls of different song birds have convergently been influenced by predator pressure; species not at all closely related use calls of a pitch which serve the warning function very well and render it difficult for vertebrate predators of a certain size to locate the sound. Haartman (1957) stressed the convergences in the nest-luring displays of widely different hole-nesting birds. Tinbergen (1959) suggested that appeasement postures, such as facing-away, bending the head down, and pointing the bill up, might well have arisen independently in many birds, perhaps even in other tetrapods. This phenomenon is really a special case of a more general phenomenon, that of "indirect effects of selection."

*Indirect selection effects.*—A number of cases are known where the nature of signaling behavior must be assumed to have been determined indirectly by selection pressure in other contexts. The similarity of nest-showing displays in birds pointed out by Haartman (1957) and a very similar phenomenon in fish, reported by Nyman (1953) illustrate how nest-site selection has developed alongside such signaling behavior. E. Cullen (1957) has convincingly argued that facing-away by the chick of the kittiwake and the possession of a black neck band, which presumably makes the movement more conspicuous and thereby effective, are related to the chick's immobility, which is required by the nesting habitat—narrow ledges on perpendicular cliffs. This habitat selection, extreme among gulls, is itself an anti-predator response. The kittiwake also fights in a way different from other gulls; unlike them it does not peck at opponents from above but pecks (as befits a cliff-nester) equally well at opponents higher on the cliff and at those below. As a consequence, it has lost the threat posture derived from this fighting pattern: the upright-threat posture which is found in all other gulls.

Hoogland, Morris, and Tinbergen (1957) have shown that the spines of three-spined sticklebacks protect them against certain predators. This has allowed this species to invade more open habitats than related species and also to develop extremely bright epigamic colors.

Nesting habitat seems also to have favored, in some as yet not understood way, the development of black coloration in the frontal part of fish of different families breeding under stones (Wickler, 1959).

## Conclusion

It is hardly necessary to point out once more that the evidence on which this sketch of the evolution of signaling systems is based is extremely patchy. Not only do gaps exist in the descriptive evidence, but systematic studies of the origin of signals are still extremely rare. Our comparative knowledge is sufficient only in a few cases to allow even sketchy descriptions of the courses ritualization must have taken, and it is at the moment quite impossible to say what changes in behavior mechanisms must have taken place (Blest, 1961). There is also a great need for experimental studies of the effects and the functions of signals. Moreover, it is becoming increasingly clear that a full understanding of the bewildering diversity and elaboration of signals will not be possible unless the full life histories of animals are studied; this alone can enable us to see the undoubtedly very numerous and intricate indirect effects of selection. Finally, our ideas about the nature of the selection pressures involved will remain conjectural until we are able to apply selection under controlled conditions. This field of research is, however, being opened up (see, e.g., Koopman, 1950; Knight *et al.*, 1956; and Pearce, 1960). However, even though experimental attack is the ultimate aim, observational studies as reviewed here shall have to provide the hypotheses and the program of work to be done.

### REFERENCES

Allen, A. A. 1934. Sex rhythm in the ruffed grouse (*Bonasa umbellus* L.) and other birds. *Auk*, **51**: 180–99.

Andrew, R. J. 1956. Some remarks on behaviour in conflict situations, with special reference to *Emberiza spp. Brit. J. Anim. Behav.*, **4**: 41–5.

Bastock, M., Morris, D., and Moynihan, M. 1953. Some comments on conflict and thwarting in animals. *Behaviour*, **6**: 66–84.

Blair, W. F. 1958. Mating calls in the speciation of anuran amphibians. *Amer. Nat.*, **92**: 27–51.

Blest, A. D. 1957. The function of eyespot patterns in the Lepidoptera. *Behaviour*, **11**: 209–55.

———. 1961. The concept of "ritualization." *In:* W. H. Thorpe and O. L. Zangwill (eds.), *Current Problems in Animal Behaviour*, pp. 102–24. Cambridge: Cambridge University Press.

Bergman, G. 1946. Der Steinwälzer *Arenaria i. interpres* (L.) in seiner Beziehung zur Umwelt. *Acta Zool. Fenn.*, **57**: 1–151.

Busnel, R. G., and Giban, J. (eds.). 1960. *Colloque sur la protection acoustique des cultures et autres moyens d'effarouchement des oiseaux.* Paris: I.N.R.A.

Butenandt, A. 1959. Wirkstoffe des Insektenreiches. *Naturwissenschaften,* **46**: 461.

Cinat-Tomson, H. 1926. Die geschlechtliche Zuchtwahl beim Wellensittich. *Biol. Zbl.*, **46**: 543.

Crane, J. 1955. Imaginal behavior of a Trinidad butterfly, *Heliconius erato hydara* Hewison, with special reference to the social use of color. *Zoologica,* **40**: 167–96.

———. 1957. Basic patterns of display in fiddler crabs (Ocypodidae, Genus *Uca*). *Zoologica,* **52**: 69–82.

Cullen, E. 1957. Adaptations in the kittiwake to cliff-nesting. *Ibis,* **99**: 275–302.

Cullen, J. M. 1960. The aerial display of the Arctic tern and other species. *Ardea,* **48**: 1–39.

Davenport, D. 1958. Observations on the symbiosis of the sea anemone *Stoichactis* and the pomacentrid fish *Amphiprion percula. Biol. Bull.*, **115**: 397–410.

Dieterlein, F. 1959. Das Verhalten des syrischen Goldhamsters (*Mesocricetus auratus* Waterhouse). *Z. Tierpsychol.*, **9**: 169–207.

Drees, O. 1952. Untersuchungen über die angeborenen Verhaltensweisen bei Springspinnen (Salticidae). *Z. Tierpsychol.*, **9**: 169–207.

Fabricius, E. 1951. Zur Ethologie junger Anatiden. *Acta Zool. Fenn.*, **68**: 1–175.

Fiedler, W. 1957. Beobachtungen zum Markierungsverhalten einiger Säugetiere. *Z. Säugetierk.*, **22**: 57–76.

Frings, H., *et al.* Recorded calls of herring gulls as repellants and attractants. *Science,* **121**: 340–41.

Frisch, K. von. 1954. *The Dancing Bees.* London: Methuen & Co.

Frisch, K. von, and Jander, R. 1957. Über den Schwänzeltanz der Bienen. *Z. Vergl. Physiol.*, **40**: 239–63.

Goethe, F. 1955. Beobachtungen bei der Aufzucht junger Silbermöwen. *Z. Tierpsychol.*, **12**: 402–33.

Goodwin, D. 1952. A comparative study of the voice and some aspects of behaviour in two Old-World jays. *Behaviour,* **4**: 293–316.

Haartman, L. von. 1957. Adaptations in hole-nesting birds. *Evolution,* **11**: 339–48.

Hediger, H. 1949. Säugetierterritorien und ihre Markierung. *Bijdragen tot de Dierkunde,* **28**: 172–84.

Hinde, R. A. 1955. A comparative study of the courtship of certain finches (*Fringillidae*). *Ibis,* **97**: 706–46.

———. 1959. Behaviour and speciation in birds and lower vertebrates. *Biol. Rev.*, **34**: 85–128.

Hinde, R. A., and Tinbergen, N. 1958. The comparative study of species-specific behaviour. *In:* A. Roe and G. G. Simpson (eds.), *Behavior and Evolution,* pp. 251–68. New Haven: Yale University Press.

Hoogland, R., Morris, D., and Tinbergen, N. 1957. The spines of sticklebacks (*Gasterosteus* and *Pygosteus*) as means of defence against predators (*Perca* and *Esox*). *Behaviour,* **10:** 205–36.

Huxley, J. S. 1923. Courtship activities of the red-throated diver (*Colymbus stellatus* Pontopp.); together with a discussion on the evolution of courtship in birds. *J. Linnaean Soc.,* **35:** 253–93.

Iersel, J. J. A. van, and Bol, A. 1958. Preening of two tern species. A study of displacement activities. *Behaviour,* **13:** 1–88.

Immelmann, K. 1959. Experimentelle Untersuchungen über die biologische Bedeutung artspezifischer Merkmale beim Zebrafinken (*Taeniopygia castanotis* Gould). *Zool. Jahrb. Syst.,* **86:** 437–592.

Keenleyside, M. H. A. 1955. Some aspects of the schooling behaviour of fish. *Behaviour,* **8:** 183–248.

Knight, G. R., Robertson, A., and Waddington, C. 1956. Selection for sexual isolation within a species. *Evolution,* **10:** 14–22.

Koopman, K. K. 1950. Natural selection for reproductive isolation between *Drosophila pseudo-obscura* and *D. persimilis. Evolution,* **4:** 135–48.

Lanyon, W. E. 1957. The comparative biology of the meadowlarks (*Sturnella*) in Wisconsin. Publ. No. 1. Cambridge, Mass.: Nuttal Ornithological Club.

Lack, D. 1947. *Darwin's Finches.* Cambridge: Cambridge University Press.

Lehrman, D. S., Brody, P. N., and Wortis, R. P. 1961. The presence of the mate and of nesting material as stimuli for the development of incubation behavior and for gonadotropin secretion in the ring dove (*Streptopelia risoria*). *Endocrinology,* **68:** 507–16.

Lindauer, M. 1955. Schwarmbienen auf Wohnungsuche. *Z. Vergl. Physiol.,* **37:** 263–324.

Lorenz, K. 1935. Der Kumpan in der Umwelt des Vogels. *J. Ornithol.,* **83:** 137–213; 289–413.

———. 1937. The companion in the bird's world. *Auk,* **54:** 245–73.

———. 1941. Vergleichende Bewegungsstudien an Anatinen. *J. Ornithol.,* **84:** 194–294.

———. 1958. The evolution of behavior. *Sci. Amer.,* **199:** 67–83.

Magnus, D. 1958. Experimentelle Untersuchungen zur Bionomie und Ethologie des Kaisermantels *Argynnis paphia* L. *Z. Tierpsychol.,* **15:** 397–426.

Makkink, G. F. 1936. An attempt at an ethogram of the European avocet (*Recurvirostra avosetta* L.) with ethological and psychological remarks. *Ardea,* **25:** 1–62.

Manley, G. H. 1960. The swoop and soar performance of the black-headed gull, *Larus ridibundus* L. *Ardea,* **48:** 37–51.

Marler, P. 1955. The characteristics of some animal calls. *Nature,* **176:** 6.

———. 1957. Specific distinctness in the communication signals of birds. *Behaviour,* **11:** 13–40.

———. 1959. Developments in the study of animal communication. *In:*

P. R. Bell (ed.), *Darwin's Biological Work,* pp. 150–207. Cambridge: Cambridge University Press.

Mayr, E. 1960. The emergence of evolutionary novelties. *In:* Sol Tax (ed.), *Evolution after Darwin,* I, pp. 349–80. Chicago: University of Chicago Press.

Morris, D. 1955. The feather postures of birds and the problem of the origin of social signals. *Behaviour,* **9:** 75–114.

———. 1956. The functions and causation of courtship ceremonies. *In:* M. Autuori, *et al., Coll. Internat. de l'instinct,* pp. 261–86. Paris: Masson et Cie.

———. 1957. "Typical intensity" and its relation to the problem of ritualisation. *Behaviour,* **11:** 1–13.

Nicolai, J. J. 1959. Familientradition in der Gesangsentwicklung des Gimpels (*Pyrrhula pyrrhula* L.). *J. Ornithol.,* **100:** 39–47.

Nyman, K. J. 1953. Observations on the behaviour of *Gobius microps. Acta Soc. Fauna Flora Fenn.,* **69:** 1–11.

Pearce, S. 1960. An experimental study of sexual isolation with the species *Drosophila melanogaster. Animal Behaviour,* **8:** 232–33.

Perdeck, A. C. 1958. The isolating value of specific song patterns in two sibling species of grasshoppers (*Chorthippus brunneus* Thunb. and *C. biguttulus* L.). *Behaviour,* **12:** 1–76.

Pelkwijk, J. ter, and Tinbergen, N. 1937. Eine reizbiologische Analyse einiger Verhaltensweisen von *Gasterosteus aculeatus* L. *Z. Tierpsychol.,* **1:** 193–201.

Rowell, C. H. F. 1961. Displacement grooming in the chaffinch. *Animal Behaviour,* **9:** 38–63.

Sauer, F. 1954. Die Entwicklung der Lautäusserungen vom Ei ab schalldicht gehaltener Dorngrasmucken (*Sylvia c. communis* Latham) im Vergleich mits später isolierten und mit wildlebenden Artgenossen. *Z. Tierpsychol.,* **11:** 10–93.

Schenkel, R. 1956, 1958. Zur Deutung der Balzleistingen einiger Phasianiden und Tetraoiniden, I and II. *Ornithol. Beobacht.,* **53:** 182–201; **55:** 65–95.

Schiller, C. H. (ed.). 1957. *Instinctive Behavior.* New York: International Universities Press.

Selous, E. 1901. *Bird Watching.* London: Constable & Co.

———. 1933. *Evolution of Habit in Birds.* London: Constable & Co.

Simmons, K. E. L. 1955. The nature of the predator-reactions of waders towards humans; with special reference to the role of the aggressive, escape and brooding drives. *Behaviour,* **8:** 85–112.

Spurway, H., and Haldane, J. B. S. 1953. The comparative ethology of vertebrate breathing, I. *Behaviour,* **6:** 8–35.

Thorpe, W. H. 1958. The learning of song-patterns by birds, with special reference to the song of the chaffinch, *Fringilla coelebs. Ibis,* **100:** 535–71.

Tinbergen, N. 1940. Die Übersprungbewegung. *Z. Tierpsychol.,* **4:** 1–40.

———. 1948. Social releasers and the experimental method required for their study. *Wilson Bull.*, **60:** 6–51.

———. 1952. "Derived" activities; their causation, biological significance, origin and emancipation during evolution. *Quart. Rev. Biol.*, **27:** 1–32.

———. 1953. *Social Behaviour in Animals.* London: Methuen & Co.

———. 1954. The origin and evolution of courtship and threat display. *In: Evolution as a Process*, pp. 233–51. London: Allen S. Unwin.

———. 1959. Comparative studies of the behaviour of gulls (*Laridae*); a progress report. *Behaviour*, **15:** 1–70.

WEIDMANN, U. 1958. Verhaltensstudien an der Stockente (*Anas platyrhynchos* L.), II. *Z. Tierpsychol.*, **15:** 277–300.

WICKLER, W. 1959. Die Oekologische Anpassung als ethologisches Problem. *Naturwissenschaften*, **46:** 505–509.

J. P. SCOTT

# 9

# The Effects of Early Experience on Social Behavior and Organization

## Introduction

This is a subject which has great practical and clinical interest. Most of our theories concerning education and mental health stress the importance of early, as opposed to later, experience, but only recently have we begun to accumulate a solid observational and experimental basis for these theories. In this chapter, I am going to review these findings in relation to the general principles which they illustrate, using certain species and types of experiments as examples rather than trying to review all of the literature. Many of the species chosen for study belong to the families Canidae (wolves and dogs) and Anatidae (ducks and geese), two highly social families of mammals and birds having close associations with man. The two major types of experimentation are cross-fostering between species and rearing in isolation. Both of these produce major changes in the social environment of the developing young. However, before we can analyze the effects of these experiments, we must understand the nature of the normal behavior which they affect.

## Some Fundamental Concepts of Social Behavior and Organization

The most general function of behavior is adaptation to environmental change. All animals which are capable of behavior show such adap-

J. P. Scott is chairman, Division of Behavior Studies, at Roscoe B. Jackson Memorial Laboratory, Bar Harbor, Maine.

tations, and each species achieves adaptations in particular ways. These are called patterns of behavior, and individuals within a species usually have a choice of several. Even in the lower animals there is some variability of behavior, so that if one pattern of behavior does not result in adaptation, another one may be substituted. Such patterns may be extremely complex in higher animals, but any given species still has characteristic ways of behaving.

For example, two male dogs or wolves approach each other for the first time. They walk in stiff-legged gaits with tails erect and tail tips moving slowly from side to side. They may touch noses and then each starts nosing the base of the other's tail, apparently investigating the other's odor. Or, they may suddenly spring at each other, snapping, snarling, and growling. If a full scale fight develops, one of the animals will eventually get hurt and attempt to run away. If he is caught, he will roll on his back, pawing with his legs, and snapping and yelping whenever the other animal attempts to bite.

These are some of the patterns of agonistic behavior which are typical of the Canidae. In part, they are the result of structure, for the only way in which one dog can hurt another is with his teeth. They also result from learning, an experienced animal being ordinarily superior to one which has never fought before (Scott, 1950).

The most important types of adaptive functions can be classified

TABLE 1

General Types of Adaptive Behavior

| Type of Behavior | Synonyms | Definition |
|---|---|---|
| Investigative | Exploratory | Investigating social, biological and physical environment |
| Shelter-seeking | Contactual, Aggregative, Comfort-seeking | Seeking out and coming to rest in most favorable part of environment |
| Ingestive | | Eating and drinking |
| Eliminative | | Behavior associated with urination and defecation |
| Sexual | | Courtship and mating behavior |
| Epimeletic | Care-giving | Giving care and attention |
| Et-epimeletic | Care-soliciting | Calling and signaling for care and attention |
| Allelomimetic | Mimesis, Contagious behavior, Mutual imitation | Doing the same thing, with some degree of mutual stimulation |
| Agonistic | Conflict, Aggression and defense | Any behavior associated with conflict, including fighting, escape, and freezing |

under nine different headings, as seen in Table 1. This provides an outline for systematic description of a species and also gives a list of the possible ways in which social behavior can be affected by experience.

A second fundamental concept of social behavior and organization is that of the social relationship, defined as regular and predictable behavior exhibited between two individuals, usually of the same species. One example of such a social relationship is dominance-subordination, described in Chapter 1. This relationship develops gradually in dogs. When young puppies first begin to play with each other they roll over and over, pawing and biting. Sometimes this develops into an actual fight, ending when one puppy gives up and runs away. Thereafter the relationship is likely to develop further along the same lines, so that when we give the two puppies a single bone the dominant puppy takes it and the subordinate one waits until the other is through. This is not the only sort of relationship which may be developed. In some cases, the first puppy to get the bone is allowed to keep it and the other one waits. In still others, the question of dominance is never really settled, and some sort of fight is likely to break out in any competitive situation. Each relationship is thus made up of certain patterns of behavior, i.e., dominance-subordination may include several patterns of agonistic behavior.

TABLE 2

IMPORTANT TYPES OF SOCIAL RELATIONSHIPS IN VERTEBRATES

| Relationship | Example | Type of Social Behavior Involved |
|---|---|---|
| Sexual | Mated pair in doves | Sexual |
| Care-dependency | Parental care in birds, mammals | Epimeletic plus ingestive, shelter-seeking, eliminative, etc. |
| Mutual care | Mutual grooming, primates | Epimeletic |
| Leader-follower | Mother and offspring in sheep | Allelomimetic |
| Dominance-subordination | Peck order, chickens | Agonistic |
| Mutual avoidance | Wild deermice | Agonistic |

There are many other kinds of social relationships, some of which are outlined in Table 2. An important general principle is that patterns of behavior and social relationships are both developed. The young puppy does not come into the world with either. The capacities for attacking and running away develop slowly as the puppy grows and are modified by learning as well as heredity. In short, the social behavior of an individual changes with time and cannot be considered a constant except in an artificial sense.

## Basic Experiments with Early Social Experience

### CROSS-FOSTERING BETWEEN SPECIES

This is part of the normal experience of domestic dogs, which are routinely taken from their mothers as young puppies and reared by human beings. We might therefore expect that the behavior of dogs would show the maximum effects of such social experience. One way to find out what these effects are is to observe the behavior of dogs toward one another and compare it with their behavior toward human beings. We placed a number of young dogs in large, open fields where they could live together and could be observed while entirely separated from people. The idea was to find out if domestic dogs had developed, either through experience or through selection, special patterns of behavior toward human beings different from their responses toward dogs. Actually, these dogs interacted with the same patterns of behavior that they normally exhibit toward human masters. Furthermore, when dog behavior is compared with that of wolves, the wild ancestors of dogs, the same sort of behavior patterns are seen. The chief effect of selection has been to exaggerate or partially suppress various kinds of behavior in different breeds (Scott, 1950).

Changing the social environment under which they have been raised for hundreds of generations has not changed dogs into people. They still act in the same ways that wolves act; the chief differences are that these reactions are now given to human beings as well as to other dogs and that dogs are much more variable than their wild relatives. In addition, these behavior patterns can be suppressed or exaggerated by training. For example, dogs which react sexually to human beings are punished and do not show the same type of behavior again.

Dogs are usually not removed from their mothers until they have had some experience with their own kind, nor are they kept entirely away from other dogs. However, such separation experiments can be carried out quite easily with birds hatched in an incubator. One of the commonly observed effects upon birds reared entirely by humans is that they not only transfer their usual behavior patterns to human beings but that they are relatively unresponsive to their own kind.

Whenever there is a choice between another bird and a human being, they choose the person, and such preferences often render them ineffectual in mating and parental behavior (Lorenz, 1935). Hand-reared male turkey poults respond sexually to the human hand more often than to models of turkeys, whereas the reverse is true of normally reared birds (Schein and Hale, 1959).

We can conclude that the major effect of this kind of early social experience is to alter social relationships rather than social behavior patterns. In the many experiments of this type, the adopted animal does not take up the social behavior patterns of human beings but develops social relationships with people on the basis of its own native behavior patterns.

### THE EFFECTS OF REARING IN SOCIAL ISOLATION

It is difficult to rear an animal completely apart from all social contacts. However, this is not necessary in order to produce some very striking effects. By the time a young puppy is three weeks of age, it can feed itself quite successfully. If such a puppy is placed in a box and fed so that it never sees or has direct contact with human beings or other dogs thereafter, its behavior becomes very different from that of an animal reared with its own kind. For example, fox terrier puppies, which come from a breed selected for fighting abilities, usually begin to fight with each other at about seven weeks of age if left with their mothers. This becomes so severe that they often have to be separated to prevent their killing each other, although a maximum of two or three can usually get along. If puppies of the same strain of fox terriers are raised in boxes until they are sixteen weeks of age before they are introduced to normally raised fox terriers, they do not attack the other animals and, if attacked, fight back unsuccessfully. They live in peace with fox terriers brought up like themselves and almost invariably get the worst of any encounter with a normally reared animal (Fisher, 1955; Fuller, 1962).

Here we see the general principle that one of the major effects of isolation is to prevent the development of social relationships. The isolated fox terriers had nothing to fight with at the age when fighting normally begins. Some of them bit their food dishes and played with them, throwing them around. As adults they showed a great deal of this type of behavior, suggesting that they may have attempted to develop social relationships with inanimate objects rather than with their own kind. The experience of isolation thus prevents the development of social relationships. Whether such puppies could still develop normal relationships, given special training, has not yet been determined, but they probably retain some capacity for adaptation. Compared with normal animals, they are greatly handicapped in their capacities for developing social relationships.

## The Descriptive Study of Social Development

Before we can experiment intelligently with the development of behavior, we must learn something about its normal course. The best way to do this is to make systematic daily observations of young animals in a reasonably normal environment. This can usually be done more satisfactorily in a laboratory than in the field, because many animals rear their young in holes or nests where observation is difficult. In such cases, the laboratory environment is often no more restricted than that of animals reared in the wild.

Observations should be made systematically and should cover major changes in the capacities of the animals. These include changes in the function of the sense organs, changes in motor capacities such as locomotion, grasping, etc., changes in psychological capacities such as learning and discrimination, and, finally, changes in the major patterns of social behavior. Motion pictures should be made at regu-

TABLE 3

Periods of Development in the Puppy and Song Sparrow

| Puppy | | | Song Sparrow | | |
|---|---|---|---|---|---|
| Name | Age (weeks) | Begins with | Name | Age (days) | Begins with |
| Neonatal | 0–2 | Birth, nursing | Stage 1—Nestling | 0–4 | Hatching, gaping |
| Transition | 2–3 | Eyes open | Stage 2 | 5–6 | Eyes open |
| Socialization | 3–10 | Startle to sound | Stage 3 | 7–9 | Cowering—first fear reactions |
| | | | Stage 4—Fledgling | 10–16 | Leaving nest—first flight |
| | | | Stage 5 | 17–28 | Full flight |
| Juvenile | 10— | Final weaning | Stage 6—Juvenile | 29— | Independent feeding |

The periods of development described by Nice (1943) for the song sparrow roughly correspond to similar periods in the puppy. In both species, the young are born in an immature state, require intense parental care and feeding, and go through much the same stages before becoming independent. Development is much more rapid in the bird, although in small mammals such as mice a similar state of maturity is reached within the same time.

lar intervals, as this makes it possible to compare different ages simultaneously.

When all the data has been gathered, it can be assembled in the form of a developmental graph or map. This usually shows certain times when a great many changes occur, dividing development into periods (Table 3). From the viewpoint of social behavior, the most important periods are those associated with important changes in social relationships. The following example illustrates the nature of these periods in the development of the puppy.

### NEONATAL PERIOD

The newborn puppy is blind and deaf. Its methods of locomotion is a slow crawl while throwing the head from side to side. When it touches anything warm and soft, it attempts to suck; and when it locates a nipple, it sucks vigorously, pushing the breast with its forepaws and pushing forward with the hind paws. If the mother is away, the puppies' movements bring them into contact with one another and they soon settle down in a heap. The mother, coming near them, pokes them with her nose and this contact sets off searching movements and eventual nursing. As they nurse, the puppies touch one another and thus keep themselves stimulated. The mother also licks the puppies in the ano-genital region with her tongue, which sets off reflexes of urination and defecation. She licks up their leavings, and in this way the nest is kept clean. In spite of being deaf, the puppies vocalize loudly in response to pain, hunger, or cold. This usually stimulates the mother to care for the puppy (Scott and Marston, 1950).

Puppies in this period thus show behavior highly adapted for neonatal life and quite different from adult forms of social behavior. Nursing is a neonatal form of ingestive behavior, and reflexive elimination is likewise a neonatal characteristic. The searching movements are a neonatal form of investigative behavior. There is no indication either that the newborn puppies are able to locate the mother through scent or that the performance improves with practice. Learning is extremely slow and uncertain compared to later stages (Cornwell and Fuller, 1961; Stanley *et al.*, 1963).

### TRANSITION PERIOD

The first change in behavior takes place on the average at about two weeks of age, when the eyes open. During the following week a series of rapid changes take place. At approximately three weeks of age, the ears open, and the animals first give a startle response to sound. When the eyes open, the puppies begin to crawl backwards as well as forwards. By the end of the week they can stand and walk. At the same time, their first teeth appear, and they are able to eat and drink by themselves. At the beginning of the week it requires eighty trials to produce a stable conditioned avoidance response; at its end, the same effect can be produced in less than ten trials (Klyavina *et al.*, 1958). They begin to eliminate by themselves outside the nest by this time. They also begin to take notice of dogs and people at a distance and will approach and investigate them. Some playful biting and pawing, or agonistic behavior, occurs. In short, the puppies have gone through a rapid transition from neonatal to adult capacities in the sense organs, in motor and psychological development, and in many social behavior patterns.

### PERIOD OF SOCIALIZATION

All of these changes mean that the puppy is now capable of developing true social relationships. During the next few weeks, the puppy shows a typical response to any new individual. For the first few minutes, he is afraid. If no harm comes, he approaches the stranger, sniffing and wagging the tail. Thereafter he is highly responsive, and the further development of the social relationship depends upon the behavior of the stranger. Among themselves, puppies spend a great deal of time in playful fighting: rolling over and over each other, pawing, biting, and clasping. By thirty days of age, the litter moves as a group on many occasions, thus showing allelomimetic behavior. By nine weeks, recognizable sexual behavior can be seen. The only type of social behavior which the puppies do not show in some form is the care-giving behavior typical of parents.

At the same time, the et-epimeletic, or care-soliciting, behavior of whining and yelping has declined a great deal and only appears under certain circumstances. The puppy will still yelp when cold, hungry, or in pain but in many cases can now adapt to these stimuli by its own activities. Other circumstances which now bring out yelping behavior are being caught, being left alone, and particularly being isolated in a strange place.

The puppy's brain is also developing rapidly, as indicated by the electroencephalograph. The newborn pup shows almost no brain waves at all and no distinction between sleeping and waking states. By three weeks of age, the brain waves are beginning to appear, and there is a difference in the sleeping and wakeful conditions. By seven weeks of age, the brain waves are essentially similar to those of the adult (Charles and Fuller, 1956). Since the major brain waves are associated with the development of vision in human beings, it is probable that these changes indicate the development of the visual cortex.

### JUVENILE PERIOD

At seven weeks of age there is an important change in behavior, as this is the time when a mother begins to wean her pups completely. They are still unable to find food for themselves and in a wild state would be dependent on food brought to them by the mother. While they are still nursing, the mother may vomit food for them, and this will be continued for some time after.

The puppies stay closely around their den or nest box for some time, not moving far away until approximately twelve weeks of age. Then they begin investigating nearby areas. They probably would be incapable of any real hunting until after the second teeth appear, beginning about four months of age. Studies of wolves indicate that these animals begin to hunt at about six months of age (Murie, 1944).

### FURTHER DEVELOPMENT

The next major change in social relationships occurs at sexual maturity, when the puppies are first capable of developing sexual relationships. This may occur at anywhere from six months to a year of age in most dog breeds, although the wild Canidae usually do not breed until the end of the second year. The birth of the young initiates a new type of social relationship and the parental period of development.

### COMPARATIVE STUDY OF MAMMALIAN DEVELOPMENT

The dog is an animal which is born in an immature state and which goes through a relatively long period of maternal care. We can now contrast its development with that of the sheep, which is born in a much more mature state and receives relatively little maternal care.

Like the dog, the newborn lamb begins life by nursing, but it begins the transition to the adult type of ingestive behavior earlier, at about ten days of age, when it first begins to graze. Unlike the dog, the period of nursing is a long one, lasting at least three months and requiring close association between mother and lamb.

Most of the transitions that occur in the dog's third week of development take place before birth in the sheep. A young lamb is born with fully developed sense organs and adopts the adult method of locomotion a few minutes after birth. It is probably capable of learning immediately after birth, and experiments show that it can be conditioned within a few days (Moore, 1958).

The formation of social relationships also begins at birth or shortly after. The newborn lamb rises on wobbly legs and approaches the nearest sheep, poking its nose under her body and eventually reaching the nipple where it begins to suck. However, only its own mother will allow it to nurse, and other sheep butt it away (Collias, 1956). This limits its first social relationship to one animal, the mother, and the young lamb follows her closely during the first few days. Likewise, the mother never moves more than a few feet away from the lamb. By ten days of age, the young lamb begins the formation of other social relationships with animals of its own age, following them and showing various playful forms of adult social behavior, such as sexual mounting and butting.

All during the nursing period, the mother and lamb are closely associated, and if they are separated they try to get together again. Usually the mother baas and the lamb comes running, but if the lamb is caught and held in some way, it will call and the mother will come to it. In any frightening situation, the lamb runs to the mother and stays close by her side. In this way a strong habit of following is built up. As a result, in adult sheep the oldest females tend to become leaders of the flock, followed by their respective offspring (Scott, 1945).

The periods of development are not as clear-cut as in the dog. However, changes in social relationships take place at ten days (when there is a shift toward solid food and a tendency to form new social relationships with animals other than the mother), at the time of final weaning, and again in the autumn when the lambs become sexually mature.

Thus there is a recognizable neonatal period and a transition period, but the period of socialization overlaps both of them. We can conclude that these periods can vary greatly in form in different animals and that we are basically dealing not with a time sequence but with certain processes, particularly the processes of transition from the neonatal to the adult forms of social behavior and the process of socialization. These processes do not occur in the same sequence in different species but differ in relation to the type of social organization typical of the adult animals (Fig. 9.1).

In the sheep, the earliest and strongest social relationship is developed with the mother and forms the foundation for leadership in

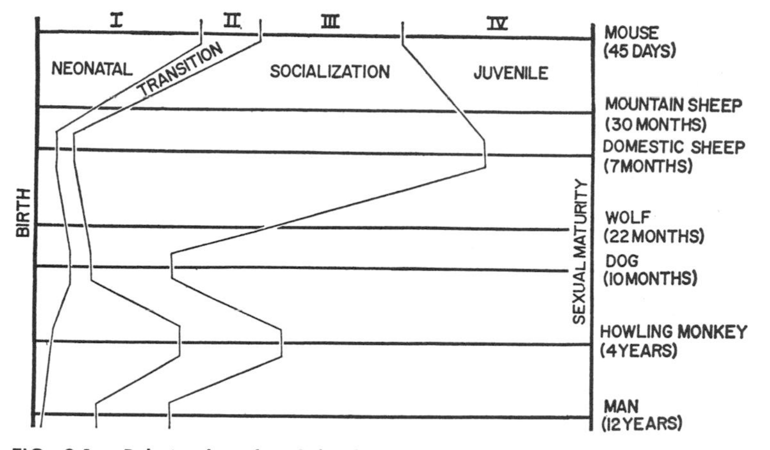


FIG. 9.1. Relative lengths of the first four periods of development in various mammals are shown in this graph. The length of each period is shown as a proportion of the total time required for the attainment of sexual maturity in females of that species as given in the figures on the right. Estimates of the period of socialization are based on the time of weaning. Actually, the period of primary socialization in man begins immediately after the neonatal period (see text). (After J. P. Scott and M. Marston, *J. Genet. Pyschol.* 7 [1950].)

the flock. In the dog, development is such that the strongest social relationships are developed with the litter mates, forming the basis for pack organization of adults. Wolves show a similar type of development, and relationships with the parents are comparatively weak because the adults spend so much time hunting during the period of socialization of the young.

We may also conclude that there is a critical period for the process of socialization (Scott and Marston, 1950; Scott, 1962). In an animal like the sheep, this occurs almost immediately after birth, but in the dog, it comes considerably later in development. In both species, there is a time when it is easy to form a social relationship with the young animal. The period is a critical one, because it determines with which particular members of its own species the animal will develop strong social relationships.

## Development of Social Behavior in Birds

In contrast to mammals, birds are primarily animals whose life is spent in the air and whose bodies are adapted for flight. This affects their physical structure, their body covering, their sense organs, and their social behavior (see Chaps. 2 and 4 for general discussion of bird adaptation).

An essential part of adaptation for flying is the possession of feathers, which provide a strong and lightweight means of increasing the surface of the wings and insulating against heat loss. This automatically limits the use of the tactile sense, since any extensive contact ruffles the feathers. With their great rapidity of movement, birds have developed the eyes as the dominant sense organs. Odor is of little importance, and hearing is not highly developed.

The important motor organs are the wings, which means that any manipulation must be done with the feet or beak. This latter organ is again connected with flight, being a lightweight replacement of teeth. Furthermore, birds tend to be relatively small in size compared to mammals.

Another major difference is the fact that all birds lay eggs, which must be kept at body temperature for incubation. This affects both parental behavior and the development of behavior in the young. Compared with mammals, their development is extraordinarily rapid. Smaller birds hatch within twelve to fourteen days and may leave the nest and fly ten days after hatching (Nice, 1943).

Nice (1943, 1950) has carefully described stages of development in passerine birds, particularly the song sparrow (Table 3, p. 236). The newly hatched bird is blind and almost naked. In the first stage, its behavior consists almost entirely of gaping so that it can be fed by the parents. In the second stage, the eyes open and the bird begins to stand up and do some preening. In the third stage, there is a rapid acquisition of adult motor patterns, such as stretching the wings, scratching the head and shaking the body. The fourth stage begins with the most important change—leaving the nest at ten days of age. Along with this come the behavior patterns of hopping, flying, walking, and sleeping in the adult position. Independent feeding does not occur for another two days. The fifth stage begins at seventeen days with the attainment of skillful flight and ends with the complete independence of the young bird and breaking of the parent-offspring relationship at approximately four weeks of age.

Comparing this with the development of mammals, we can see that these birds have a "newly hatched" period comparable to the neonatal period, in which all behavior is of an infantile type and primarily concerned with nutrition. The transition to the adult form of locomotion begins at the second stage, reaching its most important point at Stage 4 when independent flight begins. We could therefore call the second and third stages transition periods. Nice has found that the birds can be socialized to people (i.e., form their first social relationship with the human handler) if they are taken from the nest just before the fourth stage, when they normally leave the nest. It is difficult to tell when the stage of socialization begins, but it is probably as early as the fourth stage. The end of the fifth stage, when the birds achieve complete independence, corresponds to the process of weaning in a mammal and to the beginning of the juvenile period.

Lorenz (1935) reports that the jackdaw, a bird belonging to the crow family (*Corvidae*), will become strongly attached to human beings if hand rearing begins just at the time when the bird is ready for flight. Adoption at a later age produces a less complete social relationship. It seems likely that the process of socialization begins very close to the time of flight, in birds which are hatched immaturely and fly immediately after leaving the nest. This receives some support from observations of development in the cowbird, which does not become attached to the foster parents. However, it is also possible that the formation of social relationships in these parasitic birds is governed entirely by heredity.

## DEVELOPMENT OF BEHAVIOR IN GEESE AND DUCKS

These birds are hatched in a much more mature state than those described above. The newly hatched gosling has its eyes open, can walk, and even swim. The parent birds may assist the young in finding food but do not feed them. In spite of this maturity, young birds go through a relatively long period before they develop flight feathers and make the transition to the adult pattern of locomotion. By this time they are nearly full sized.

The newly hatched goslings will follow any large, moving object.

This is usually the parent bird but may be a boat which comes near the nest of the wild bird, or, in the case of birds hatched in an incubator, may be the person of an experimenter. Once the young goslings have followed an individual for a few hours, it is almost impossible to get them to transfer to another; and they consequently develop their further social relationships with the bird or person to whom they first became attached. This process of primary socialization in birds has been called "imprinting" (Lorenz, 1935), implying that the young birds are in some way permanently impressed or molded by their early social experience. This phenomenon was discussed in some aspects in Chapters 4, 6, and 8, described more extensively in Chapter 7, and will be dealt with here in relation to the formation of the social bond.

The process of socialization normally begins soon after hatching. Most of the transition from infantile to adult forms of behavior has already taken place in the egg, but the most important transition, to flight, takes place at a much later period, at six weeks or more of age. In contrast to the perching birds, in which the transition to flight occurs almost simultaneously with the beginning of socialization, the primary social relationships are formed long before. This is correlated with a different kind of social organization as adults. In geese, ducks, and similar birds, the families stay together at least until the fall migration, with considerable evidence of leadership. Family groups in the perching birds generally break up immediately, and no leadership is apparent.

### EXPERIMENTS WITH IMPRINTING

Most of these have been done with ducks of various wild and tame varieties and with domestic chickens, whose eggs are more easily obtained (Hess, 1959). The latter have a type of development somewhat similar to that of ducks and geese and pass through the same early period in which primary socialization or imprinting takes place.

Many of these experiments are concerned with the nature of the stimulation which produces imprinting and are done with animated models which bear some degree of resemblance to the parent animals. Fabricius (1951) found that the ducklings would follow a model of almost any shape, including square boxes, and of almost any size, except very small objects. The effectiveness of the model can be greatly increased if some noise accompanies it. This does not have to be the call of the parent bird, almost any monotonous noise being effective, and means that the ducklings respond to a very general sort of stimulation. Once having responded, they rapidly learn to discriminate between the special model and others. Imprinting is thus limited to the parent species by the circumstances of hatching and maternal care rather than any innate response to special stimuli.

Fabricius and Boyd (1954), Hess (1959), and others have found that the young ducklings show the maximum tendency to follow between one and two days of age. Younger and older animals followed much less readily. This confirms the existence of a short critical period for the process of imprinting (Fig. 9.2).

Fabricius also found that the young ducklings showed escape as well as following responses to the model. As the animals grew older, they showed more and more escape responses, indicating that it is the development of fearful behavior which terminates the critical period. In terms of natural behavior, the young ducklings would normally be brooded and protected by their parents during the first few days of life and would have an opportunity to follow only them. Following the parent birds would keep the young ones away from others; but if strange animals did come near, the escape response would prevent the formation of an association. This negative limitation on primary socialization is quite similar to that found in dogs and may be a general one among vertebrate animals (Hess, 1959).

These experiments with models have emphasized the effect of the first social interaction upon the young animal, neglecting the fact that there is a corresponding effect upon the parents. The normal situation is one in which there is interaction between the behavior of parents and young, forming a social relationship involving both animals. Such experiments do show, however, that social relationships of an imperfect sort can be formed with animated or even motionless models, as well as with animals of different species.

## The Results of Early Social Experience

King (1958) has reviewed the factors which must be studied for a complete analysis of the effects of early experience. The age at which the experience is given is extremely important, as is its duration. Furthermore, simple chronological age means little except in relation to the developmental periods of the species studied.

In a satisfactory experiment it must be first established that there

FIG. 9.2. Imprinting in ducks has been analytically studied in the apparatus *above*. The decoy "mother" moves around the runway producing the "glock" call. The duckling to be imprinted is placed on the outside runway and allowed to follow the decoy. Among the factors analyzed using this apparatus was the relative sensitivity of animals exposed at different ages to a standard period of imprinting and tested later for their responses. As shown in the chart (*below*) these ducklings were most effectively imprinted when from thirteen to sixteen hours old and were but little affected by the same experience after twenty-four hours. (After E. Hess, *in* E. Bliss (ed.), *Roots of Behavior,* 1962.)

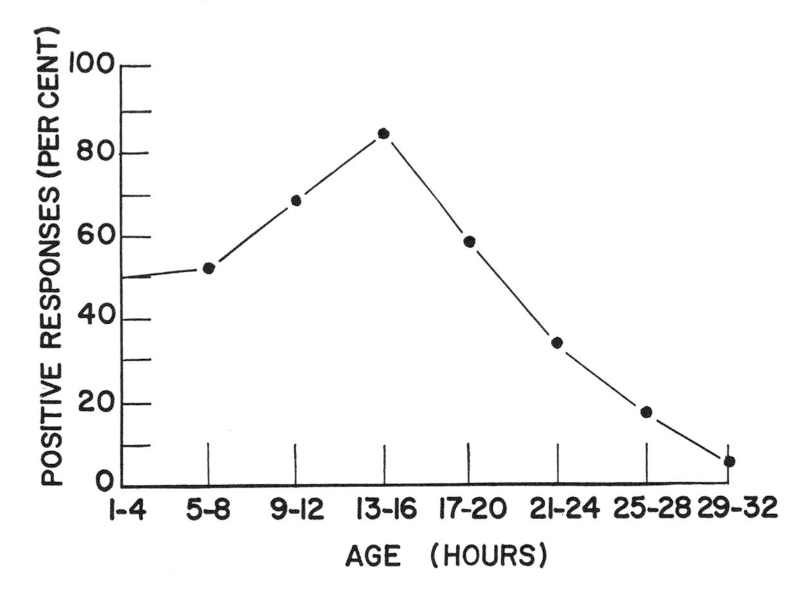


is some immediate effect of the early experience. Second, it must be shown by testing at a later date that there is a long-range effect. The age at testing should be primarily related to the periods of development rather than chronological age. Finally, it is not sufficient to show merely that an early experience has a later effect. The test must determine whether the effect was important and whether it will persist if the animal is given a chance to readjust its behavior.

As I said above, the principal types of experiments have been the fostering of young animals by a different species, usually man, and comparing animals raised in isolation with those reared in normal social groups. These treatments have relatively little effect on the patterns of social behavior peculiar to a species but do produce major effects on social relationships. Early social experience thus has a profound effect on social organization.

### SEXUAL RELATIONSHIPS

Hand-reared birds of many species tend to transfer their courtship and mating behavior to human beings, resulting in interference with the formation of normal sexual relationships and mating bonds. However, in most experiments there has been little attempt to differentiate between the situations which result in lasting effects and those which might permit some readjustment. In the case of hand-reared mammals, the usual effect is for the hand-reared individual to respond both to human beings and to receptive females of the proper species. Results of this kind are often seen in animals reared as pets.

The other type of experiment (rearing animals in social isolation and comparing them with animals reared in groups) also has effects upon sexual relationships. Beach (1942) found that male rats reared alone after weaning showed more readiness to mate than those reared in groups. Sexual behavior is not otherwise affected, and the explanation seems to be that the young rats reared together form habits of passive inhibition or active play which interfere with sexual behavior later. However, this interference is not serious enough to prevent effective mating. It should be pointed out that the play of young rats is largely aggressive rather than sexual. The sexual responses of females are not affected by isolation (Stone, 1926).

Guinea pigs are much more precocious animals than rats, being able to walk, follow their mothers, and even eat a little solid food at birth. Valenstein and his associates (1955) took male guinea pigs and divided them into two groups. The isolated ones were left with their mothers but away from littermates up to twenty-five days of age, then completely isolated until seventy-seven days, at which time their sexual behavior was tested. The social males were left with their mother and siblings for twenty-five days, then each placed with three to five females up to seventy-seven days. All were then exposed to females in heat, and the socially experienced animals were found to be much more effective in their sexual behavior. The differences were more striking if the isolated males were taken from their mothers at ten days of age, and similar results were obtained for females. In this species of mammal the result of early social experience is to facilitate sexual behavior.

In still another species, mice, isolation has no effect on adult sexual behavior of males (King, 1956). The explanation of the different results in various species of mammals seems to lie in differences in the type of early social experience. In many immature mammals, such as lambs, goats, dogs, and monkeys, a considerable portion of play behavior is sexual in nature. Dr. Albert Pawlowski reared a female puppy at our laboratory apart from dogs but in contact with human beings. When she came into estrus as an adult, she was introduced to several male dogs. Many of her responses were inappropriate, e.g., rolling over in a playful defensive posture when approached by the male. However, an experienced male was able to hold her in the proper position and complete mating.

At the Yerkes Primate Laboratories young chimpanzees are taken from the mother at birth and reared by human beings in a nursery for the first year or two of life. Nissen (1954) placed mixed groups of these young animals together and left them until just before puberty. None of the inexperienced animals were able to complete mating. However, experienced animals were able to mate with members of the group, the males being more effective than the females. Thereafter, the experienced young animals were able to mate with one another. Nissen concluded that in these animals mating behavior has to be learned from older individuals.

Harlow (1963) has obtained equally dramatic results with young rhesus monkeys reared with dummy mothers. Both males and females are completely unable to mate with one another, and Harlow has never been able to induce mating in the males, even with experienced females. In a few cases the females were impregnated by experienced males. The whole difficulty can be averted if the young animals are allowed a few hours of social play during a critical period. Young monkeys normally indulge in a good deal of playful sexual behavior during which the correct adult patterns of behavior are developed.

Thus there are two demonstrated effects of early social experience upon sexual behavior. One is the formation of habits which interfere with later adult behavior, as in male rats. The other and more common effect is practice in playful sexual behavior which facilitates adult behavior. Which of these is most important depends upon the species and the age at which the experience occurs.

### DOMINANCE-SUBORDINATION RELATIONSHIPS

Early social experience has a pronounced effect on agonistic behavior. Young animals of many species normally do not fight with the particular adults or young with whom they become socialized. As they grow older, they may form a dominance-subordination relationship with them and continue to live in relative peace, at the same time beginning to fight fiercely against strange animals similar to those with which they were brought up.

Rearing in isolation modifies the development of agonistic behavior in many ways, depending on the species concerned. Guhl (1958) found that young cockerels reared in a group did not begin to form a peck order until they were approximately six weeks of age. However, if he reared them in isolation and put them together at thirty-one days of age, they began to form a peck order immediately.

Among mice, animals reared in the same litter live peaceably together long after they reach the adult stage, and they may never fight at all. If they are exposed to strange animals, however, they begin to fight with them at about thirty-two days of age, when puberty and male hormone production begins. King (1957) studied the effect of early social experience on this tendency to fight strangers. He kept mice with the mother until twenty days of age, which is about a week before the normal weaning time and the first age at which the young animals can survive successfully without the mother. Thereafter he kept some males together for various periods and isolated others. He found that the animals reared with other males attacked strangers much sooner than did the isolated ones. He also found that at least ten days of social experience were necessary to produce the effect and that the same experience as an adult produced no effect at all. There is obviously a critical developmental period for controlling the appearance of this effect. The explanation seems to lie in the fact that isolated males investigate a stranger very cautiously and may never get started fighting at all. Males with social experience seem to recognize the stranger as an enemy immediately and waste no time in starting to fight.

Puppies during the period of socialization exhibit all sorts of playful behavior, including playful fighting. In the more aggressive breeds, this eventually leads to real fights and the formation of definite dominance-subordination relationships between members of a litter. King (1954) tested litters of adult basenjis and cocker spaniels which had been reared in this way, exposing them to strange animals of the same two breeds. He found that they exhibited most aggression against strangers most like themselves; that is, cocker spaniel females tended to attack strange cocker females most severely, and basenji males exhibited most hostility against males of their own breed. The explanation seems to be that the animals recognize the behavior of those similar to themselves but are not inhibited by dominance relationships when the like individuals are strangers.

Wire-haired fox terriers are highly aggressive even as young puppies. By seven weeks of age, they may attack one another so severely as to produce physical injury or even death, although they can form dominance relationships between small groups of two or three pups (Fuller, 1953). As stated above, Fisher (1955) and Fuller (1962) raised such puppies in complete isolation from three to sixteen weeks of age and found that they did not fight when placed together. In this case isolation can completely prevent the development of the dominance-subordination relationship.

As in sexual relationships, there are two effects of early social experience. Socialization can inhibit fighting by the processes of passive inhibition and the early formation of dominance relationships (Scott, 1958). On the other hand, early experience involving mild or playful fighting behavior facilitates fighting against strangers.

### CARE-DEPENDENCY RELATIONSHIPS

What is the effect of fostering or isolation upon the behavior of an animal which later became a parent and attempts to rear its own offspring? Little work has been done along these lines because the experimental animals must be reared to adulthood. One might expect that a fostered animal would have some difficulty in recognizing its own young. A female sheep was raised apart from the flock and returned at the age of ten days (Scott, 1945). It never became a part of the flock but accepted a male at breeding time and subsequently gave birth to a lamb. The lamb developed the tendency to follow the flock and frequently became separated from its mother. Unlike the normal mothers, the parent showed little anxiety at the absence of her offspring, although the lamb frequently became anxious when the mother was missing. This suggests that such animals might make poor parents.

Among rats, part of the pattern of maternal behavior is building a nest for the young. Riess (1950) reported that rats which were raised on powdered food and given no bedding when young did not build nests for their offspring even though bedding material was provided for this. Eibl-Eibesfeldt (1955) repeated this experiment but only examined nest building of single rats without offspring. He found that if he lowered the temperature, most of the rats without previous nest-building experience built them immediately. Likewise, inexperienced rats were much more likely to build nests when given crepe paper than when provided with straw. It looks as if the nest-building pattern of behavior is organized largely by heredity and that experience chiefly has the effect of making animals respond to a lesser degree of stimulation and a wider variety of materials.

Much more dramatic effects result from isolation of young primates. Harlow (1963) finds that female rhesus monkeys reared with dummy mothers and deprived of contact with other monkeys later become extremely poor mothers. Whereas the normal mother continually holds its baby to her breast and responds to its cries with great solicitude, these orphan mothers are either indifferent to their young or actively reject them, pushing them away and beating them when they cry. In this species, maternal behavior is organized by the combined effects of heredity and early social experience.

### LEADER-FOLLOWER RELATIONSHIPS

Lorenz' experiments with hand-reared graylag geese showed dramatically how the leadership relationship could be transferred to human beings. As adults, the flock would fly after him as he rode a bicycle or swim after him as he swam in the river.

Similarly, the hand rearing of young lambs transfers the leader-follower relationship to the human handler. If permitted, the young lamb will follow its foster parents everywhere. The relationship between the lamb and its own kind is at first completely disrupted, although a young male forms some tendency to follow through sexual attraction.

In summary, the results of early social experience upon the later development of social relationships can be extremely drastic under certain circumstances. Isolation prevents the normal development of such social relationships. Depending on the species, subsequent development may be completely or only partially disrupted. The normal process of primary socialization may have two conflicting effects. One of these effects is the formation of strong habits of behaving in an immature fashion toward the individuals involved, thus inhibiting the adult behavior toward these same individuals. This plays an important part in the control of agonistic behavior between members of a permanent social group, and it may in some species tend to prevent sexual behavior between closely related animals. The other is the facilitation of development of adult sexual and parental patterns of behavior, this being particularly important in various species of primates.

## The Effects of Early Social Experience on Human Behavior

It is impossible to perform drastic experiments with young human infants, such as rearing them in isolation or fostering them on different species. There have been a few criminal cases of child abuse involving complete isolation, and various semi-mythical accounts of children reared by animals (Ogburn, 1959). None of these has produced any valid evidence, and we can learn most by a careful study of normal development in human infants (Scott, 1963).

The human infant goes through a neonatal period lasting a month or six weeks during which time its behavior is almost entirely devoted to the infantile mode of feeding. It gives responses to light and sound, but studies with the electroencephalograph show no alpha rhythm, indicating that the visual cortex is undeveloped. Probably the human neonate actually sees very little.

The first change in behavior is the appearance of the ability to form stable conditioned reflexes rapidly. Shortly thereafter, the infant begins to smile at human faces, and the alpha waves appear in the EEG. From this time up until about six months of age, the child will smile at any stranger. After this a fear of strangers and strange situations begins to appear. It looks as if the time from six weeks to six months is the period of primary socialization, although further socialization can take place later, and strangers can form a relationship with the child if they persist in their contacts.

The most important transition processes begin at approximately seven and a half months of age, at the time of the appearance of the first teeth and development of the ability to crawl. By fifteen months of age, the child has the first four grinding teeth and is beginning to walk. This may properly be called a transition period. A further transition takes place in communication, since fifteen months is the time when a child begins to talk. By two years or shortly afterward, he is beginning to talk in short sentences, has a complete set of baby teeth, and can run as well as walk. This is the time when the child first becomes moderately independent of the mother, and thus this point in development corresponds to the age of final weaning and the beginning of a juvenile period in other mammals. All this means that the period of primary socialization precedes the major periods of transition in a human infant and that the first relationships will be developed with the mother or whoever cares for the child.

From a standpoint of practical child care, we can say first of all that the human infant at birth is highly protected from psychological damage, both by the circumstances of normal parental care and by the sensory, psychological and motor immaturity of the infant. Second, the period from six weeks to six or seven months is one in which new social relationships are developed with great ease. We could predict that relationships would be transferred from one person to another with little disturbance of the child, and this seems to be the case with regard to adoptions. We can also conclude that disturbance of primary social relationships during the next period might have much more serious results, and the evidence which Bowlby (1951) and others have gathered from children sent to hospitals at this age indicates serious emotional disturbance, although it may not necessarily have a lasting effect.

Finally, we can say that in attempting to predict the outcome of any social experience, we should keep in mind the kinds of factors which have been discovered experimentally in other animals. The developmental period is very important, and almost equally important is the genetic nature of the child affected. Some individuals are highly sensitive and others relatively insensitive. Early social experience may result either in the improvement of social relationships by practice or the opposite, inhibition of adult behavior by formation of conflicting habits. Finally, we must recognize the ability of a child to learn independently. Two children, given the same experience, may draw opposite conclusions from it. Thus there is no ironclad determination of adult behavior by early experience. We can, however, hope that in the future we may be able to use early experience more intelligently than we now do to promote mutual tolerance and stable conditions of mental health.

## REFERENCES


BEACH, F. A. 1942. Comparison of copulatory behavior of male rats raised in isolation, cohabitation, and segregation. *J. Genet. Psychol.*, **60**: 121–36.

BOWLBY, J. 1951. *Maternal Care and Mental Health.* Geneva: World Health Organization.

CHARLES, M. S., and FULLER, J. L. 1956. Developmental study of the electroencephalogram of the dog. *Electroenceph. Clin. Neurophysiol.*, **8**: 645–52.

COLLIAS, N. E. 1956. The analysis of socialization in sheep and goats. *Ecology*, **37**: 228–39.

CORNWELL, A. C., and FULLER, J. L. 1961. Conditioned responses in young puppies. *J. Comp. Physiol. Psychol.*, **54**: 13–15.

EIBL-EIBESFELDT, I. 1955. Angeborenes und Erworbenes im Nestbauverhalten der Wanderatte. *Naturwissenschaften*, **23**: 633–34.

FABRICIUS, E. 1951. Zur Ethologie junger Anatiden. *Acta Zool. Fenn.*, **68**: 1–178.

FABRICIUS, E., and BOYD, H. 1954. Experiments on the following reaction of ducklings. *Wildfowl Trust Annual Report*, **6**: 84–89.

FISHER, A. E. 1955. "The effects of differential early treatment on the social and exploratory behavior of puppies." Ph.D. thesis, Penn State University.

FULLER, J. L. 1953. Cross-sectional and longitudinal studies of adjustive behavior in dogs. *Ann. N.Y. Acad. Sci.*, **56**: 214–24.

———. 1963. Programed life histories and socialization of the dog. *Proc. Int. Cong. Psychiat.* Montreal (in press).

GUHL, A. M. 1958. The development of social organization in the domestic chick. *Animal Behaviour*, **6**: 92–111.

HARLOW, H. F., HARLOW, M. K., and HANSEN, F. W. 1963. The maternal affectional system of rhesus monkeys. *In:* H. L. RHEINGOLD (ed.), *Maternal Behavior in Mammals.* New York: John Wiley & Sons, Inc.

HESS, E. H. 1959. The relationship between imprinting and motivation. *Nebraska Symposium on Motivation*, **7**: 44–77.

KING, J. A. 1954. Closed social groups among domestic dogs. *Proc. Amer. Philo. Soc.*, **98**: 327–36.

———. 1956. Sexual behavior of C57BL/10 mice and its relation to early social experience. *J. Genet. Psychol.*, **88**: 223–29.

———. 1957. Relationships between early social experience and adult aggressive behavior in inbred mice. *J. Genet. Psychol.*, **90**: 151–66.

———. 1958. Parameters relevant to determining the effects of early experience upon the adult behavior of animals. *Psychol. Bull.,* **55:** 46–58.

KLYAVINA, M. P., *et al.* 1958. On the speed of formation of conditioned reflexes in dogs in ontogenesis. *Acad. Sci. U.S.S.R. J. Higher Nervous Activity,* **8:** 929–36.

LORENZ, K. 1935. Der Kumpan in der Umwelt des Vogels. *J. Ornith.,* **83:** 137–213, 289–413.

MOORE, A. U. 1958. Conditioning and stress in the newborn lamb and kid. *In:* W. H. GANTT (ed.), *Physiological Bases of Psychiatry.* Springfield, Ill.: C. C Thomas.

MURIE, A. 1944. *The Wolves of Mt. McKinley.* (U.S.D.I. Fauna Series No. 5.) Washington, D.C.: U.S. Government Printing Office.

NICE, M. M. 1943. Studies in the life history of the song sparrow, II. *Trans. Linnaean Soc.* (N.Y.), **6:** 1–328.

———. 1950. Development of a redwing (*Agelaius phoenicius*). *Wilson Bull.,* **62:** 87–93.

NISSEN, H. 1954. Development of sexual behavior in chimpanzees. *In:* W. C. YOUNG (ed.), *Genetic, psychological and hormonal factors in the establishment and maintenance of sexual behavior in mammals.* Lawrence: University of Kansas. (Mimeograph.)

OGBURN, W. F. 1959. The wolf boy of Agra. *Amer. J. Sociol.,* **64:** 449–54.

RIESS, B. F. 1950. The isolation of factors of learning and native behavior in field and laboratory studies. *Ann. N.Y. Acad. Sci.* **51:** 1093–1102.

SCHEIN, M. W., and HALE, E. B. 1959. The effect of early social experience on male sexual behaviour of androgen treated turkeys. *Animal Behaviour,* **7:** 189–200.

SCOTT, J. P. 1945. Social behavior, organization and leadership in a small flock of domestic sheep. *Comp. Psychol. Monogr.* No. 96, **18(4):** 1–29.

———. 1950. The social behavior of dogs and wolves. *Ann. N.Y. Acad. Sci.,* **51:** 1009–21.

———. 1958. *Aggression.* Chicago: University of Chicago Press.

———. 1963. The process of primary socialization in canine and human infants. *Child Develop. Monogr.,* **28(1):** 1–47.

———. 1962. Critical periods in behavioral development. *Science,* **138:** 949–58.

SCOTT, J. P., and MARSTON, M. 1950. Critical periods affecting the development of normal and mal-adjustive social behavior of puppies. *J. Genet. Psychol.,* **77:** 25–60.

STANLEY, W. C., *et al.* 1963. Conditioning in the neonatal puppy. *J. Comp. Physiol. Psychol.,* **56:** 211–14.

STONE, C. P. 1926. The initial copulatory response of female rats reared in isolation from the age of twenty days to age of puberty. *J. Comp. Psychol.,* **6:** 73–83.

VALENSTEIN, E. S., RISS, W., and YOUNG, W. C. 1955. Experiential and genetic factors in the organization of sexual behavior in male guinea pigs. *J. Comp. Physiol. Psychol.,* **48:** 397–403.

WILLIAM ETKIN

# 10

# Types of Social Organization in Birds and Mammals

## Introduction

The concept of adaptation implies not only that all the important characteristics of the organism are suited to the needs of the organism in relation to its environment but also that these characteristics fit with each other to make for efficiently operating wholes. For example, the eye of a given vertebrate not only must be an efficient organ for accomplishing visual needs but it must be efficient in precisely those ways which the activities of that organism require and not in other ways or in higher degree than required. The focusing capacities of the eye vary enormously in different vertebrates, according to the different ecological requirements of the organism. Fish living in water necessarily have little range of vision and little need for much change in focus in their eyes. Correspondingly, we find the focusing mechanism in fish generally poorly developed or absent. Ground-living forest dwellers require, and have, greater capacity for accommodation; arboreal and plains-living animals have still more, and certain birds have the most extensive apparatus. Not only does the range of accommodation vary appropriately in these differing animals, but in each type the speed of focusing varies in correlation with the rate of locomotion of the animal. Although some mammals have keen vision and considerable depth of focus, their speed of focusing is slow in comparison with the focusing mechanisms of hawks and other birds of prey. These have accommodation mechanisms which employ rapidly contracting striated muscle, in contrast to the slow acting smooth muscle found in the mammalian eye. The hawk must necessarily keep its prey in accurate focus until the moment of contact while pouncing upon it at a speed of some two hundred miles per hour.

In a similar way the behavior system of an organism must not only be efficient but must correlate with the entire life of the organism. Survival value depends upon the over-all balance of many factors. The integration of behaviors is as significant as the usefulness of the behavior considered in isolation.

It must not be thought, however, that integration implies that all the forms of behavior are necessarily harmonious. They may, in fact, clash and, by opposing, keep each other in check. The conflicts which arise in territorial behavior and the means of their resolution illustrate this. Defense of the territory necessitates aggression against other members of the species. If carried too far, this interferes with acceptance of the mate. The level of aggression that is attained in a given species depends upon the balance of such factors as the role of territoriality, extent of sexual dimorphism, the nature of the courtship, role of the male in parental care, population pressure, danger from predators, etc. In Chapter 8 Tinbergen provided another example by pointing to the limitations imposed upon the development of conspicuousness in signaling devices by the dangers of exposing the animals to predators. To understand the integration of factors in social organization, we must think of each characteristic of behavior in quantitative as well as qualitative terms. In this chapter we propose to describe how the various elements of the behavior systems interrelate in each of several selected animals to produce a type of social organization that constitutes an effective mode of environmental adaptation.

Animal life is nothing if not variable. For almost any general statement that can be made about a group, some exception or apparent exception can be found. We can deal here only with what we regard as the generality of the interrelations representative of the principle of correlation and make little or no attempt to consider the variants.

Since this chapter reviews by animal type behaviors discussed previously in other contexts, it is neither necessary nor practical to specify the references again. With few exceptions, therefore, no direct reference to the literature is given for these items. The most important studies upon which the discussion of each type is based are listed according to types at the end of the chapter.

## Birds

### PERCHING OR SONG BIRDS

Song birds are highly mobile, diurnal animals, dependent upon vision and hearing for sensory guidance, feeding on small animals (insects, etc.) or plant products, and subject to predation from the air (hawks) and, to a lesser extent, from ground-living animals. Their behavioral system in the reproductive phase tends to be organized around seasonally defended territories limited to one mated pair. In most species male and female are alike (monomorphic). Agonistic display usually involves stereotyped sounds and movement rather than structural characteristics. Courtship is largely limited to pre-nuptial displays and is commonly of aggressive nature. Parental duties are shared more or less equally between male and female, or at least the male is helpful in raising the young. The family unit is disrupted at the end of the reproductive season, when the young become entirely independent of their parents. In both resident and migratory species flocks are formed; often different age classes or sexes flock separately.

A little reflection will show that this catalogue of behavioral items is no mere random assortment but that the factors are interrelated and interdependent. Since the male establishes his territory before mating, the pairing-up process involves the violation of the male's territory by the female. Hence, we expect male aggression to be present in courtship. If pairing is to be successful, the aggressive responses of the male must be countered by some characteristic of the female. Morphological releasers identifying the sexes would serve this purpose, and, indeed, such are found in the red-winged blackbird and flicker. Dimorphism has, however, the disadvantage of rendering the individual more conspicuous and, therefore, subject to predation. Most song birds do not exhibit dimorphism but depend upon behavioral responses, of which the passive acceptance by the female of aggression by the male is the most commonplace. Thus the characteristic courtship of song birds emerges: aggressive actions by the male and passivity by the female—as described for the song sparrow in Chapter 4. Once pairing up has been effected, courtship practically ceases, since the pairing bond is reinforced by the fact that both animals are confined to the territory. Territoriality of the song bird type also limits competition among males to the period of establishment of the territory. Aggressive display in these circumstances is sufficiently accomplished by song, posturing, and minor structural characteristics, and there is a general absence of elements tending to make the males permanently conspicuous. This permits males as well as females to assume protective colorations consistent with the important role that both parents of these species play in

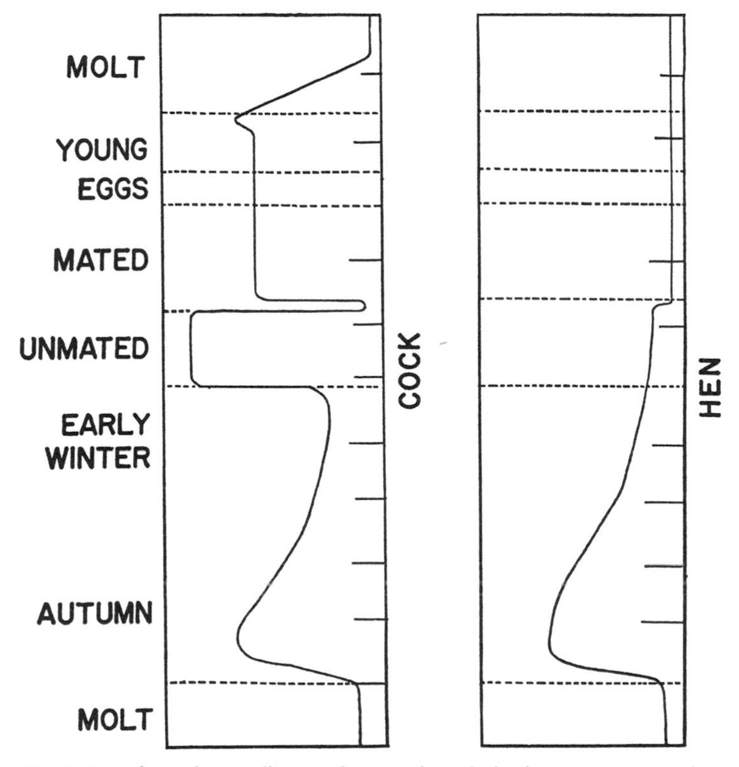


FIG. 10.1. The robin redbreast hen and cock both sing prominently in their individual territories in autumn. Thereafter the female (*right graph*) shows little song display, being especially quiet during the breeding period. Males, on the other hand, show considerable singing especially when unmated in spring. Immediately after mating and at molt they are quiet (*left graph*). (After D. Lack, *The Life of the Robin*, 1943.)

parental care. Even the danger of such an apparent exception as the red breast of robins is minimized by the fact that it is on the under side and only becomes conspicuous when exposed by the specialized postures associated with territorial defense (Fig. 10.1). The breakup of the family bonds when the young reach full flight capacities is correlated with the preparation for the fall migration. This requires a new kind of socialization without territoriality. Since feeding in flocks on widely distributed foodstuffs is non-competitive, little aggressive behavior is shown at this time.

Among song birds responses governed by simple releaser mechanisms predominate, with little dependence upon learning in evidence. For example, the responses between parents and young are largely controlled by typical kumpan relationships. Since ordinarily there are no competing young in the territory, individual recognition of young by parents or of parents by young is unnecessary. Where learning is involved, as in recognition of the nest site, it often depends upon factors other than the biological essentials of the situation. For example, the bird may choose to sit on the empty nest rather than on the eggs when these are set nearby. This simplification of social behaviors made possible by territoriality gives rise to many of the paradoxes that appear so characteristic of the study of song-bird behavior. The parent is readily "fooled" by crude substitutes for its young or eggs, as in the classic ethological experiments discussed in Chapter 7. Such birds are likewise the victims of natural "deceivers," as the parasitization of their nests by cuckoos or cowbirds illustrates. The seeming limitation of bird mentality is clearly not organically imposed and does not arise from any limitations inherent in the species' nervous system, since birds, such as crows, with different ecological requirements show considerable learning capacities. Rather, it represents a form of adaptation to their mode of life. The remark, "Birds are so stupid because they can fly," is attributed to Heinroth, one of the founders of the ethological school. By this he meant that the advantage of flight obviated many of the subtleties of escape and attack behavior that characterize mammals generally. We see that this, though valid, is only part of the story. The social life of song birds is so simplified by their type of reproductive territoriality that there is little need for complex learning activities. The evolutionary process does not call forth any greater or more complex mechanisms than those that suffice to achieve survival under the particular ecological situation.

There are many contradictory and conflicting forces acting upon any organism, and the behavior system achieved in its evolution represents a compromise or resolution of these forces—or perhaps, better said, a dynamic equilibrium between them. For example, the territoriality of song birds would seem to create a selective pressure toward dimorphism, emphasizing as it does aggressive and display potential in the males. The monogamous mating which is favored by the same territoriality, however, creates a selective pressure away from dimorphism, since conspicuousness in the male would subject him to predation and undermine his usefulness in the care of young. In general song birds tend to be monogamous (during each reproductive period) and monomorphic. Sometimes the balance of factors leans toward some conspicuousness, as in the English robins where both have the red breast and both parents defend the territory after

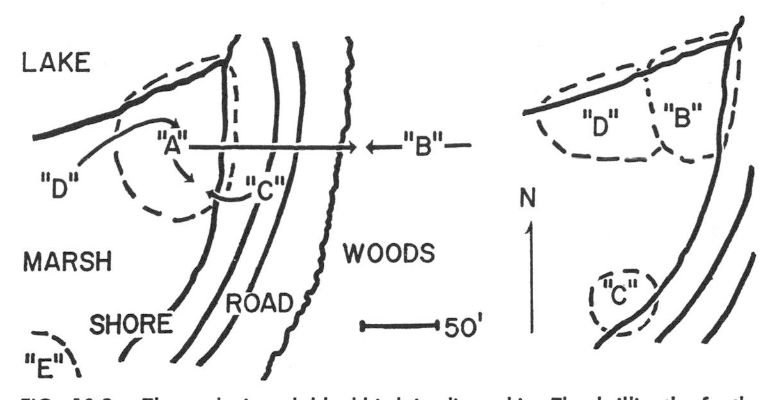


FIG. 10.2. The red-winged blackbird is dimorphic. The brilliantly feathered male holds a territory in which several females may become established. Each female has her own sub-territory around her nest, which she defends against other females. Males attract females to their own territories by courting behavior, but established females often interfere, attempting to drive the new female away. In such cases the male may or may not attack his original mate. The diagram above (*left*) shows the events in an area in Wisconsin from March 20 to April 21, 1952. Male A held male B far back in the woods. On March 30, male C arrived and challenged A. While A was fighting and driving off male C, male D quietly took over a portion of his territory. Later male A disappeared from the scene and his territory was then taken over by males D and B. Male C, after being repulsed by A, established himself in a territory nearer the shore (*right*). (After R. Nero, *Wilson Bull.*, 68 [1956].)

pairing. In the red-winged blackbird, there is considerable dimorphism in the black color and red epaulets of the male, in contrast to the mottled browns of the female. This bird has achieved this different balance favoring dimorphism, since the male often has several females in his territory and himself plays little role in parental care (Fig. 10.2, Fig. 10.3). Thus, to appreciate any one characteristic, as aggressiveness toward others of the species or dimorphism, we must see it against the background of the entire mode of life of the organism.

FIG. 10.3. In the red-winged blackbird, the male defends a territory with a "song-spread" display in which the bright-red covert feathers at the shoulders are erected and sometimes vibrate, producing a flash of color against the black background of the rest of the body (*left*). Several females may nest within one male territory, each defending her area against other females with much the same "song spread" (*right*), even though her general color is cryptic and she lacks the red epaulets. (After R. Nero, *Wilson Bull.*, 68 [1956].)

## MARINE BIRDS

Marine birds feed along the shore (gulls) or by fishing in the open sea (gannets). Their feeding ranges are therefore very extensive. Feeding is generally in flocks, which, as pointed out in Chapter 1 is more effective than non-social feeding. Breeding activity, on the other hand, is necessarily restricted to fixed shore positions and generally is characterized by the development of small breeding territories within large breeding colonies. Both parents range widely to feed, and both return to the breeding territory to take part in parental activities (Fig. 10.4). After the breeding season the animals may range far from the breeding grounds, generally in flock formations. Since these birds are usually large and strong, the predators from which they suffer are primarily those that attack the eggs and young.

The behavioral correlates of this mode of life include the following: Pairing up usually takes place in the flock outside of the territory (Fig. 10.5). The pair then takes up a very small nesting territory within the colony and both parents defend it vigorously. The parents share nearly equally in the care of the young. Post-nuptial courtship in the form of nest-relief ceremonies is often elaborately developed

(Fig. 10.6). Monomorphism prevails, male and female being so much alike as to be difficult to distinguish. Though initial parental reactions to the nest site, eggs, and young are largely governed by simple releaser mechanisms, parents soon learn to recognize their own young, and, of course, the members of a mated pair show extraordinary ability to recognize the individuality of their mates among the hundreds of highly similar animals all around. Loss of parental care

**FIG. 10.4.** Gannets in a breeding colony each occupy a small breeding territory which is vigorously defended against strangers. In the center, the nest forms somewhat of a mound of sticks, sea weed, etc., in which a single egg is found. Note the regularity of the spacing of the birds. It would appear to require a considerable feat of memory for a bird to locate its own nest, yet this is regularly accomplished with great precision. The mates then indulge in an elaborate nest-relief ceremony before the returning individual is allowed to take over the task of incubation. (Drawn from photograph.)

usually occurs abruptly at the appropriate stage of the development of the young and is often accompanied by definite rejection behavior which compels the young to undertake feeding flights on their own.

The principal difference in the mode of life that we note here in comparison to the song-bird pattern is the separation of the feeding and breeding situations. The persistence of courtship (nest-relief ceremonies) throughout the period of parental care appears to be adaptive to this situation, since the parents must leave the breeding territory to feed and return to the confusion of the breeding ground.

Correlated with this is the highly developed learning ability shown in recognition of mates, which is obviously necessary if the birds are to go from and come to their territories freely. Vigorous defense of the breeding territories is essential for protection against nest robbing. The elaborate nest-relief ceremonies appear to permit the suppression of the highly developed territory-defense reactions when the mate returns. The monomorphic characteristic, as in song birds, correlates with the similarity of the parents in relation to care of the young. Here, however, territorial defense is at closer quarters, and morphological display characteristics are even less evident than in song birds; such elaborations as song and flight displays are notably absent, as would be expected under the conditions of crowding usually obtaining at the breeding site. The fact that pairing takes place before territory establishment is correlated with the small size of the territory. Its size is usually such that the aggressive courtship of song birds would be impossible in it.

FIG. 10.5. Food-begging is a common courtship pattern in birds. The male gull (*right*—note swollen neck indicating regurgitation) often feeds the female during courtship. At this time she begs food from him with the same begging movements the young animal shows to its parents (*left*). (After N. Tinbergen, *The Herring Gull's World,* 1953.)

As has been discussed in the section on courtship and in Chapter 6, social stimuli sometimes stimulate the endocrine system through the brain-pituitary connection. The dense clustering of sea birds in their breeding colonies can be understood as an example of the effects of such stimuli, for it has been found that the correlation of activities among different members of a colony are better achieved in large colonies than in smaller, presumably because of the influence of the social stimuli provided by neighbors to one another. This, in turn, makes for better reproductive success of the colony as a whole. For example, if the young of one pair begin wandering from their nest sites while neighboring pairs are still vigorously defending territory, they are likely to be killed. The importance and delicacy with which such social factors operate is seen in instances where, as a result of outside interferences, the social interactions among colony members

FIG. 10.6. These figures illustrate the mutual courtship displays in gulls. Oblique postures of a male herring gull (*A*) at his territory shows aggressive motivation, since it is part of the attack posture. This signal repels males but attracts females much as the singing of a territorial song sparrow. When the female arrives, the mutual courtship that follows shows many displays that indicate the mixed motivation of aggression, fear, and sexual attraction. Mutual mew calling (*B*) indicates aggression since it is often given in frankly conflict situations. However, in the conflict, individuals face each other with their threats instead of assuming parallel positions close to each other as in courtship. Choking (C) likewise indicates aggression modified by the tendency to stay. In the kittiwake, choking is given only in the nest situation and thus appears to include the parental motivation. The movement itself appears as the intention movement ritualized from the regurgitation activity in feeding young or mate. Facing away (*D*) in the black-headed gull appears as aggression inhibited by fear. It appears to be derived from the intention movement of turning to flee. As a signal of low level of aggression it serves as an appeasement sign which inhibits the aggression of the mate. Thus the courtship, pre- and post-nuptial, of the gulls shows displays indicating different mixtures of motivations and appearing to be signals "correctly interpreted" by the interacting gulls. (After N. Tinbergen, *Behaviour,* 15 [1959].)

are interfered with by disturbance of the nesting birds. In such cases nest robbing may become endemic in the colony and result in a total failure of the reproductive process in the colony for that year.

### FLOCK-LIVING SURFACE-FEEDING BIRDS

There is considerable contrast between the general aspects of the breeding patterns of marine birds and those of the flock-living ground and water birds such as the pheasants, turkey, barnyard fowl, and many ducks (Fig. 10.7). Such birds generally move in flocks over

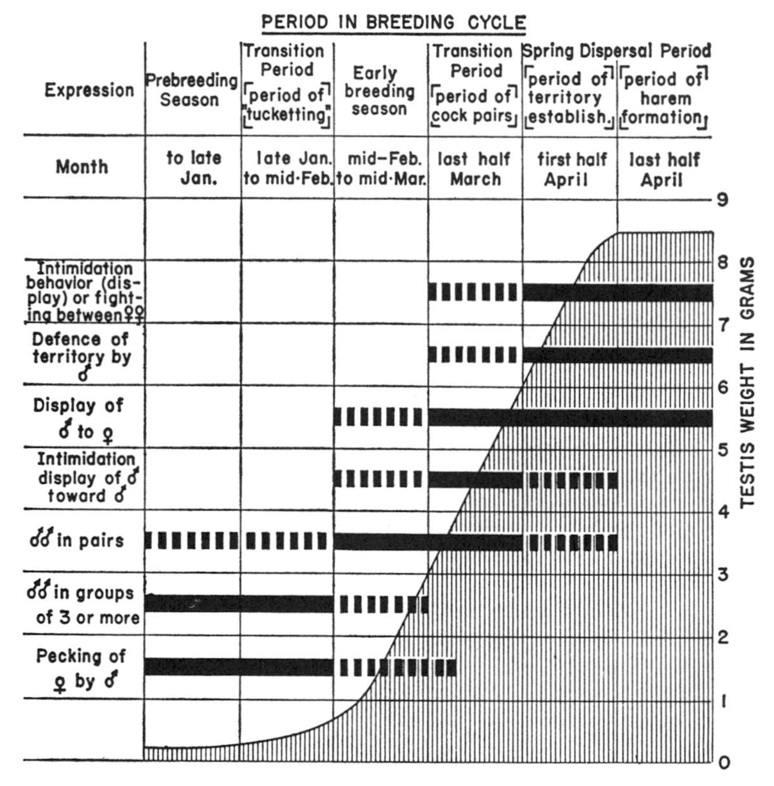


FIG. 10.7. The sequence of behavioral characteristics in pheasants during the year is shown here. During the winter the males tend to stay in unisexual flocks and show aggression toward females. At this time their testis (*shaded area*) are small (*two lowest spaces*). As spring approaches the males separate, show aggression toward other males, and display to females. In the April breeding season the males defend the territory, and the females show aggression toward each other. The testes have reached maximal development at this time. (After N. E. Collias and Taber, *The Condor*, 1951.)

large areas, feeding chiefly off the surface. The young are necessarily precocious and follow the mother and feed for themselves early, often when only a few hours old. Females and young are protectively colored, but the adult males are among the most conspicuously and gorgeously attired of all birds.

These birds also differ behaviorally from marine birds to an extreme degree. There is no reproductive territory established by the males though special mating territories may be established in some species. In many species each male forms a harem of females which he dominates and from which he excludes other males. The female usually defends the nest site. The dominance hierarchy in the flock is clear-cut, linear, and absolute in form, aggressiveness being especially conspicuous among males. Parental care is typically left almost exclusively to the females, although dominance may be modified in the direction of males calling females to food (chickens). Males act as the defenders of the flock. Responses to predators depend chiefly upon innate releasers of instinctual escape reactions. Learning ability is conspicuously developed in respect to individual recognition of members of the group—as is necessary for hierarchy. Food discrimination is also readily learned, as would be expected in general foragers (chickens). The learning of parent-young relation furnishes the paradigm of imprinting (chickens, ducks). The adaptive nature of this constellation of behavioral and structural characteristics for surface-living birds is readily seen. Flock-living is protective in preventing surprise by predators and as such is of supreme importance for surface-living, diurnal animals. This and the need for a wide foraging range preclude the limited territoriality of the song bird type. Flock-living gives prominence to the development of dominance hierarchy; and this, operating in the sphere of reproduction, places a high premium upon male dominance, since only the dominant males will succeed in breeding (Fig. 10.8). In such flocks, too, the success of reproductive behavior depends upon female responsiveness during estrus to the dominant male. These factors lead to the extreme aggressiveness and elaboration of display characteristics of the males, as noted above. Furthermore, the precocial young require no parental care from the male and only a limited amount of protectiveness from the female. The male role is thus limited largely to insemination of the females, and the expendability of the males in the harem situation, where only a few males are required for reproduction of the group, removes much of the restraint upon the development of con-

FIG. 10.8. This illustration shows the effect of dominance on breeding success of roosters. Three roosters of different strains were penned with white hens. The offsping of the Rhode Island Red are recognizable by their dark color, those of the barred Plymouth Rock are lighter and the one chick fathered by the white leghorn is white. Since the sperm of each of these varieties were shown to be equally fertile, the difference in the number of offspring shows the difference in access to the females of the roosters according to dominance order. (Redrawn after A. M. Guhl, *Scientific American*, 1960.)

spicuousness in the males which may be operative in other situations. Feathers and other ornaments have been developed to an extreme in the males of such birds (Fig. 10.9). Lek formation in some members of this group is a type of behavioral organization possible to such non-parental males and appears to take place in forms which are widely dispersed in feeding. It thus enables the females, when ready for mating, to find the males more easily.

In these animals the different roles of learning and instinct can likewise be recognized as adaptive. These are the animals in which the classic examples of absolute dominance have been so extensively studied. It can be seen that the ability to learn individual recognition is the basis for flock organization. The size of the flock in nature appears to be roughly co-ordinate with the extent of this learning ability. The closed nature of the flock formation and the extreme xenophobia shown may likewise be understood in the light of the control of aggression possible only in the hierarchically organized flock. Again the phenomenon of imprinting in parent-young relation is readily studied among such animals and is clearly part of the system of behavior that makes their survival as mobile flocks possible. On the other hand, the predator relations of these birds is much like that of song birds, and the signal system that activates their hiding-from-predator response also depends upon innate releasers.

The conclusion of our consideration of three different types of social organization among birds is that in all fundamentals, their

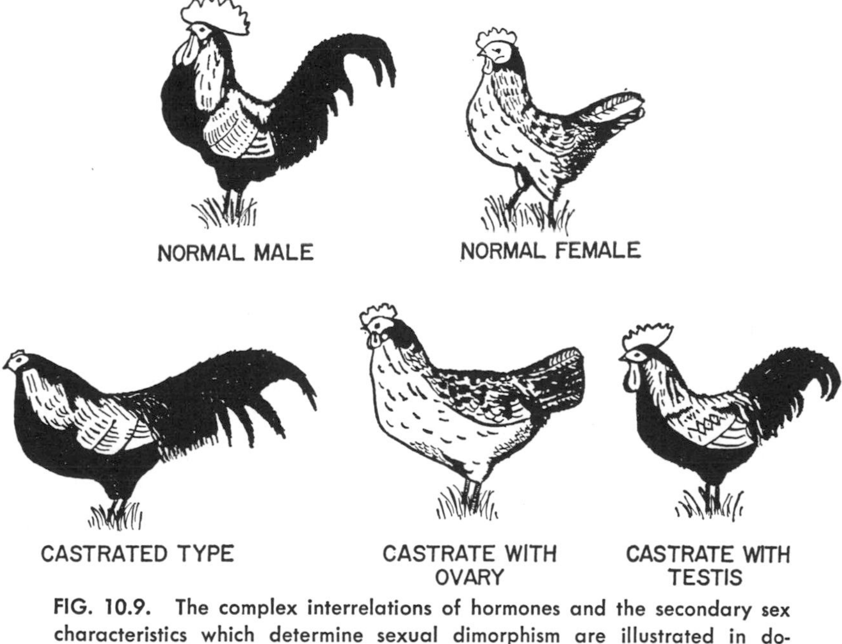


FIG. 10.9. The complex interrelations of hormones and the secondary sex characteristics which determine sexual dimorphism are illustrated in domestic fowl. The comb and wattles of the head are under the influence of male hormone. This hormone is produced in high concentration by the testes and in low concentration by the ovary. Hence the male has large head furnishings, the female small, and the castrate none. Male plumage develops when the individual matures but it does so only in the absence of female hormone. We find, then, that castrates or castrates with grafts of testes have male plumage but, with ovarian grafts, have female plumage.

behavioral factors—such as aggressiveness, hierarchy, territoriality, courtship, parental care, dimorphism, innate or learned communication—are all part of a co-ordinate complex which, in its entirety, is adaptive to the particular animal's ecology.

## Social Organizations among Mammals

### THE RAT

The wild brown rat, of which the common laboratory animal is an albino-mutant form, makes its living as a scavenger, being active at night and depending for protection upon its ability to find cover quickly. Its territoriality is essentially of the home-range type, accompanied by an intense exploratory drive. No typical defended territorial reactions are shown; however, each group tends to remain separate from others, with an uninhabited no man's land between home-range territories. Though dominance patterning can be brought out to a limited extent in artificial competitions, it is neither markedly expressed nor quickly established. Such hierarchy as is shown under experimental conditions seems based on individual habits of aggression, with little evidence of recognition of individuals as such. Most important, there is no competition between males for an estrus female, for one rat (in laboratory albinos) does not exclude another from access to an estrus female. There is slight sexual dimorphism, with males somewhat larger than females but otherwise very similar. They breed rapidly, being polyestrus during the major part of the year and, under artificial conditions, throughout the year. The young grow rapidly, reaching sexual maturity in about two months. Parental care is given only by the female and is restricted to a lactation period of about twenty-one days. Litter mates do not maintain continued association so that no "pack" organization results, but the individuals tend to remain in the same colony.

The central area of adaptation in this behavioral system appears to be that of rapid population expansion. This enables them to build up a large population to take advantage of temporarily favorable conditions, such as are afforded by seasonal food supplies. The absence of defended territory is adaptive here, avoiding as it does restrictions upon maximal reproduction. The physiological pattern of reproduction likewise favors rapid population expansion. Since gestation and lactation are of about equal length, the female nurses one litter while carrying the next. A few hours after giving birth, she again comes into heat, thus starting the next pregnancy. This reproductive pattern sometimes gets out of hand and gives rise to the familiar phenomena of mouse and rat plagues. Their intense territory-exploration drive, which is, of course, the basis of their usefulness to psychologists in maze-learning experiments, serves as part of their protective behavior. As mentioned in Chapter 7, rats kept under semi-natural conditions come to know their home territory so thoroughly they can instantly find the best and nearest shelter whenever disturbed, no matter what part of their range they may be in at the time. The highest learning capacities of rats are thus related to home range rather than to hierarchy, sex, or feeding. Their mating and parental activities seem to be largely governed by innate stimulus-response relations and stereotyped motor patterns, as might be expected from the relative simplicity of their reproductive life.

### THE RED DEER

The red-deer herd is matriarchal in organization. It is derived from the association of the young with their mothers, and thus it develops from the maternal family. Since the animals are large and relatively long-lived, three or more generations are commonly represented in the herd. The young all associate with their mother for the first three years; the males then maturing sufficiently to become sexually active at rut. They lose their association with the herd, whereas the females remain with the maternal herd. During the year the animals move through their home-range territory in a schedule dictated by seasonal conditions—the highlands in the summer and the protected valleys in the winter. The mature males live in loosely organized herds separate from the females. The latter live in herds which include the young and which are well organized. Dominance and leadership are exercised by the older females, one of which often exerts clear-cut leadership, guiding the migration, urging laggards over streams, and in other ways directing the herd's activities. The older females are constantly on the alert for signs of danger, from predators and from adverse weather changes. It is clear that they determine the movements of, protect, and guide the herd through the regular paths of migration used year after year (Fig. 10.10). In the fall rut season the male herds break up and the individuals scatter widely. They invade the female territories, round up the hinds, and keep them in

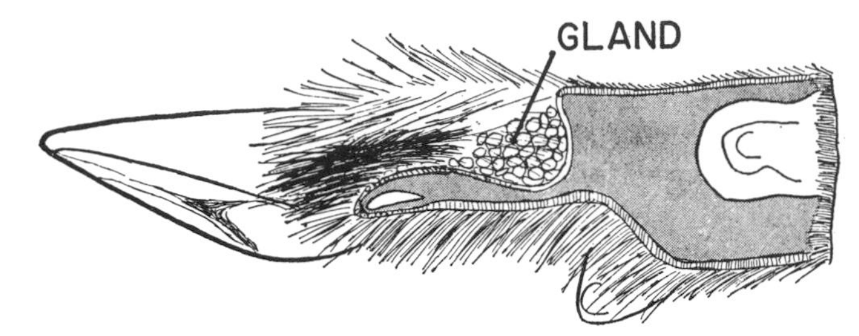


FIG. 10.10. This is a longitudinal section of the front foot of a mule deer doe showing location of interdigital gland. This gland produces secretion that lays down a scent trail by which the animals can follow each other's paths. The secretion is particularly heavy in young fawns, probably enabling the mother to trail them. Males rub their legs together thus distributing the scent of glands on the tarsal regions. They also trail the does by the scent and appear to distinguish estrus does by the smell of the urine. This is important, for in this species the males do not herd the females into harems but pursue them individually. (After J. M. Linsdale and P. Q. Tomich, *A Herd of Mule Deer,* 1953.)

harems. These males display aggressively to other males which try to approach and, as necessary, fight to exclude other males from their harem areas. As the females come into heat, they are serviced by the harem master only. This constant activity, sexual and defensive, maintained day and night by the stags usually exhausts them after about a week, and they are then forced by fresh animals to retire to the uplands, leaving their harems to the newcomers. Foals are born in the early spring, stay close to their mothers, and achieve substantial, but incomplete, growth before the first winter.

Among the behavioral correlates of this mode of life, we may note that the males show little socialization, separating from the female herds when young and showing only the loosest herd structure thereafter. In correlation with this is the strongly developed aggressive activity of the rut and the polygynous mating pattern. The role of the male is merely that of insemination, and his aggressive actions and the extreme development of aggressive potential are consistent with the dispensability of the individual male for parental functions and the polygynous structure of the society. As is to be expected in this situation, the females show little courtship activity. They tolerate the harem master but do not co-operate in his territorial-defense interests. They do, however, show clear-cut dominance hierarchy within their own herds. This hierarchy is based upon learned relations between mother and child during development. In the adult stage no aggressive action is necessary to maintain the leadership precedence. The young females are attentive to the movements and actions of the older females and appear by this attentiveness to learn the "traditions" of the herd with respect to the seasonal migratory paths, the interpretation of weather signs, and sentry activities. In contrast the male yearlings (according to Fraser Darling) are not so attentive to the actions of other herd members and do not acquire the store of knowledge of the herd. In correlation with their highly developed dimorphism in respect to aggressive potential and their pugnacity during mating, the males play no role in parental behaviors, neither protecting, assisting nor even paying any attention to the young. Except when maintaining the harem, the males show no aggressiveness toward one another. In contrast to the female herd, no hierarchy, leadership, or co-operative behavior is evident in the male herd. The intricate relations of these animals, especially the females, to one another and to the changeable weather conditions upon which their survival depends correlates with the lack of stereotyped behavior and the prominence of learning in their activities.

### THE WOLF

Wolves are pack-hunting carnivores that bring down large prey, such as deer, by co-operative action in hunting. They not only attack in a pack but carry through such pack stratagems as tiring out their prey by spelling each other off in keeping up the pursuit. The pack appears to be derived from the family unit, consisting of father and mother and offspring of both sexes and of several successive years. Apparently on occasion several such family packs may join forces, at least temporarily, to form large packs. Whether the subgroups in a large pack are blood relatives, i.e., have common familial experiences forming the basis of their tolerance for one another, or not is not known. A breeding pair generally has several dens—either natural caves or excavated holes in banks. Male and female remain together apparently for life. Adult non-breeding animals, presumably the older offspring of the senior pair, have been seen to remain associated with the pair and their young litter and to share in all activities. When burdened by pregnancy or newborn young, the female does not hunt with the pack but is fed by the male or other members of the pack who bring back part of the kill to her. After the pups are born in the spring, food is brought back to the den by all the adult wolves, sometimes in chunks and sometimes by regurgitation. Th pack appears to be a closed group, since strangers have been seen to be driven away from the group. But on other occasions, new animals were seen to join a pack. Possibly these were former members of the group. There is much play—mouthing, licking, chasing, and play-fighting—among all members of a pack at all times. Friendly greetings with tail wagging and howling are frequently exchanged between members of the pack, particularly on special occasions, as when the pack gathers together before going off on a hunt.

We note here a number of behavioral correlates to an ecology centered around pack-hunting. The female with a new litter and possibly also one late in pregnancy is unable to hunt with the pack. Co-operative behavior, marked by the bringing back of food to the den by the other adults, is adaptive in this situation. Such behavior is only one of the more dramatic examples of the co-operative behavior that marks the pack-hunting mode of life and without which this way of life would be impossible. The inter-individual behaviors of the members of the pack are marked by much social communication by play in various forms, not only among the young, but also among the adults. This would appear to be a means of constant reinforcement of the individual social interdependencies. Of course, the members of a litter, as they grow up together, engage to an even more marked degree in this social play. As in many other carnivores the young learn hunting techniques from the adults; but in this case the close socialization is not restricted, as in cats, to the mother-young relation but extends to all members of the group. Socialization by various forms of play with other members of the litter of the same generation is especially marked and appears to be the primary bond between members of the pack. There is little stereotyped behavior to suggest kumpan relations.

Among mammals the wolf type of socialization is exceptional, though not unique, in including a dominant male (father) and other non-breeding males in a co-operative group. Such a tight social group, including all members of the biological family, may be called an integrated family. Dominance hierarchy is clearly present, but its expression involves a minimum of agonistic behavior since it forms during development of the young within the family relation and remains stable throughout life. The absence of fighting is not the result of an absence of dominance, but rather, a result of the stability of the hierarchy. The extraordinary phenomenon of adults remaining within the pack without showing breeding activity, as appears to be the case in wolves, may receive its explanation from this strongly developed dominance hierarchy; for, as discussed in Chapters 4 and 5, experiment has shown a tendency in dogs and other animals for mating behavior of males to be suppressed in the presence of a dominant male (psychological castration). The system of communication between individuals within the pack is quite elaborate, as might be

FIG. 10.11. By variation in carriage and facial expression wolves communicate their social reactions to each other. In an extensive study of wolves under zoo conditions, Schenkel learned to interpret this signaling system. His inferences of the significance of some tail positions is shown above. It will be readily recognized that much of this signaling system is common to domestic dogs. (After Schenkel, *Behavior,* 1 [1948].)

expected. It involves much signaling by body, facial, and tail movements, posturing, and some vocal expression (Fig. 10.11).

### MACAQUE MONKEYS

As studied under semi-natural conditions, the Indian or rhesus monkeys and the Japanese macaques (also baboons) live in clans of some 20 to 150 animals. Their habitat is forest or semi-open country. They feed in trees or on the ground by food gathering, eating mostly plant products but also insects, bird eggs, and small animals when available. Except for nursing infants, each animal forages for itself. Usually several males are found in each clan. These males are supremely dominant within the total group and are themselves arranged in a strict linear hierarchy. Other mature males are excluded and form associated bachelor groups. The dominant males show little or no co-operative behavior toward the females, though they may take important positions when the clan is in danger. In the Japanese species, the males are reported to be somewhat more social to females and young, even playing with and caring for young to a slight extent in some groups.

The group territory is defended by the dominant males. Territory fighting is vicious, and the fighting capacity of the top males is a principal factor in determining territory size. Females in heat approach closely to the males, usually facing some adverse aggression in their initial approaches. Often a female passes from subordinate male at the beginning of estrus to the top dominant at the height of her estrus and then to a subordinate again. The young from birth associate closely with the mother, who nurses, cares for, and supervises them. Socialization depends upon these female-young contacts for its initiation but is maintained by widespread grooming activity among all members of the group and by play in youngsters. Facial expressions, postural changes, and sounds are used as signals. The necessary learning includes individual recognition, complex pathways through trees, the variety of and methods of extracting food materials, and possibly the signal system. Dominance and sexuality are closely interlocked. The sexual presentation posture is assumed as an expression of subordination and serves to halt aggression on the part of the dominant animal by channeling such aggression into mounting behavior. Mounting is thus a characteristic behavior of the dominant, as the presentation posture is of the subordinate, irrespective of the sex of the individuals involved. These behaviors are thus not strictly sexual in these animals.

The social organization of the macaque monkey is seen to emphasize the evolutionary pressure toward male aggressiveness. The dominant males exclude subordinates from breeding, and most of the offspring are thus probably fathered by the dominant males of the group. The aggressive potential of the males is further emphasized by the group territorial system, which makes group success in holding desirable territory dependent upon such potential. The exaggeration of the canine teeth, the large body size and power, and the aggressive disposition of the males is the result to be expected from these evolutionary pressures. The importance of aggressive activity in this social organization is further emphasized by the existence of bachelor males ever on the alert to invade the group.

It is difficult to see in what way intelligence and learning are favored in this ecology. It is true that the attainment of goals by roundabout paths demanded from animals with wide-ranging feeding in trees places a premium upon the type of "problem-solving" learning in which monkeys excel. The precision of binocular visual discrimination and judgment and the use of the hands in manipulation and feeding likewise favor a kind of intelligence. However, their feeding activities seem much less demanding of intelligence than the co-operative-hunting techniques of wolves.

A somewhat broader picture of the role of intelligence is emerging from the recent studies on Japanese monkeys. Food preferences and selection among the hundreds of available substances are learned by the young from other members of the group, usually the mothers. A new preference or way of handling food (i.e., washing) acquired by one adventuresome individual has been seen to spread through the group and become a culturally transmitted trait, distinctive of the group. Differences between clans in mounting activity and in the co-operativeness of males with the young are also reported and are probably culturally transmitted. Social factors, such as relationships of females to males and of young to females, are reported to influence position in the dominance hierarchy in subtle ways. The importance of these cultural factors suggests that a high level of intelligence is important in monkeys, primarily in relation to the complexity of their social life.

### HOWLING MONKEYS

Howling monkeys of the tropical rain forest live in groups much like those of macaques, except that aggressive behaviors are much less in evidence. The males show little evidence of dominance and do not even compete for the estrus females, nor do they exclude younger males from the group. The males are not indifferent to the young, although their co-operative activity is rather limited. For example, they call to attract attention to young which have gotten into difficulty by falling, although they leave the work of rescue to the females. They are not, however, aggressively negative toward young, as rhesus males often are. The males are most active in seeking pathways through the trees of the dense jungle in which they live. They direct the movements of the group by vocal signals, calling on the females and young to follow. Among the most unusual activities of the males are their howling sessions, which regularly take place in the early morning and produce a tremendous volume of sound. This is probably a signal that warns different clans of the presence of others and serves to keep the groups apart, despite the density of the plant life hiding one from the other. Several clans may cover the same general territory in their home ranges. If two clans come into close range during the day's movements, a competitive howling session determines which group yields to the other.

Compared with that of the macaque, this social system is seen to place much less emphasis upon aggressive behavior. In correlation with this, the dimorphism of the sexes is not nearly so marked as in macaques. It is possible that the permanent positions of the animals —the high tree tops of the tropical rain forest with its very dense plant growth—is basically responsible for the minimizing of fighting and the substitution of howling as a competitive mechanism. Obviously, fighting in their habitat is both dangerous and ineffective. Correlated with low aggression and lack of dimorphism is the more co-operative behavior of the males in relation to females and young. Whereas the macaque male generally ignores females and young when not actually aggressive toward them, the howling-monkey male leads or brings up the rear of the group in a protective manner. He rushes to the defense of members of the group, a behavior distinct from the defense of territory by the macaque, which ignores individuals. Altogether, the differences in behavior of the two monkeys seem adaptive to the different environmental situations of the two groups and the resulting differences in the selective forces playing upon them.

## General Principles of Behavioral Correlation

We may now attempt to summarize briefly some of the main areas of correlation that have emerged from our consideration of a few specific types of social organization in the animal world.

### DOMINANCE

Dominance hierarchy is best developed where situations of competition exist between individuals within the group, most often with respect to reproduction. Behaviors correlated with hierarchical organization include highly developed individual recognition of in-group members and antagonism to out-group individuals. In-group feeling is generally maintained by learned communicative activity, as in play and grooming, early experience being especially significant in establishing the relationship. With respect to reproduction, hierarchy favors the dominant animals in mating. It correlates with polygyny and harem formation, with dimorphism in terms of male aggressive potential, and with an absence of co-operation by the male in parental duties.

### TERRITORIALITY

Home-range territoriality correlates with the needs for security from predation and often is expressed in a highly developed exploratory drive. Defended territoriality generally correlates with the nature of the reproductive pattern. Defense is usually a male function. Group-defended territory is correlated with polygyny and exclusion of young and subordinate males from the group. The small breeding territory of a mated pair, as in gulls, is correlated with bi-parental defense of territory, monomorphism and bi-parental co-operation in raising young. Recognition of individuality in mates and in young (when these become actively mobile) is well developed. Courtship behavior continues during the parental period in the form of nest-relief ceremonies. Reproductive territory of the song-bird type has some similar correlates, including quick learning of territory, courtship which mitigates the aggressiveness of territory defense, and parental co-operation in raising young. The recognition of mate and young, however, is poor; and there is much dependence upon kumpan relations and innate releasers in social activities—including courtship, mating, and parental activities.

### REPRODUCTIVE BEHAVIOR AND DIMORPHISM

The type of courtship shown correlates with the biological characteristics of the reproductive pattern as seen in the differences between birds and mammals. The types of reproductive behavior also depend upon (*a*) differences in the roles played by the two parents, (*b*) the type of territorial defense system, and (*c*) the kind of dominance hierarchy. Dimorphism in the form either of aggressive potential or display potential interferes with co-operation by the male in the parental role and with the survival of such males. Where one parent, usually the male, is free from the requirement of co-operation in parental function, dimorphism is favored, since such males are biologically expendable in the economy of the species. Mammalian reproduction, because of the predominance of the female in relation to the young, tends to have this dimorphic nature, as does that of group-living polygynous birds.

### COMMUNICATION AND SOCIALIZATION

The nature of the stimulus-response relations found in a species is highly correlated with its ecology. Sign stimuli are conspicuous where the association of the animals is simple and not subject to too many disturbing stimuli, as in the parent-offspring interrelation in territorial birds. Imprinting characterizes ecologies such as that of ducks and ungulates where individuality of the parent-offspring relation must be maintained under conditions where many similar individuals would confuse responses based upon sign stimuli. Where socialization is of a more complex nature, as in the permanent groups of long-lived mammals, socialization is generally learned on an in-group versus out-group basis. Play and grooming, and other forms of sharing mild stimulation, by providing numerous contacts are the principal techniques for learning socialization (Fig. 10.12).

## Application of the Principles of Correlation to Man

### INTRODUCTION

The social organization of contemporary man is obviously culturally determined to such an extent that many persons doubt that biological factors play a significant role at all. The philosopher and historian Ortega y Gasset has asserted, "Man has no nature, what he has is history." That is to say, he is what his past experience has made him. However, biologists would incline to seek underlying, biologically determined tendencies even in man and to regard cultural modifica-

FIG. 10.12. The importance of social behavior in the survival of a species is beautifully illustrated in studies of prairie dogs by J. A. King. By living in colonies, they secure protection against predators by their constant watchfulness from their hills and their barking signals. Their grazing habits prevent the growth of tall grasses over a large area. They thereby encourage rapidly growing weeds upon which they feed, which in a way borders on the practice of agriculture. The members of each sub-division of a prairie-dog town constitute a coterie in which each member learns to know and accept the other members of the coterie, while rejecting outsiders. This, of course, is possible because the animals maintain constant social contact with each other by various methods of communication. Upon meeting, members frequently groom each other, play by pawing and climbing over one another (*4*) and nuzzle each other in an open mouthed kiss. Usually only one adult male (*2*) and several females with their young (*3*) constitute a coterie. Each coterie jealously guards its own territory, often less than an acre, and system of burrows. Knowledge of the territory is transmitted from generation to generation. The importance of the sense of smell in their social communication is shown not only in the friendly kissing but also by the fact that strangers turn tail toward each other and expose their anal glands for olfactory exploration (*1*). (After photographs and description by J. A. King.)

tions as being limited and guided by these. So large a question could not profitably be discussed here. However, as evolutionists, we must suppose that man at one time had a biologically determined social organization, which might be elucidated by the application of the principles of correlation developed above and applied to our knowledge of the paleontology of man. We will here attempt a brief sketch of what might have been the social organization of man at a period before culture was so predominant (protocultural man).

The question of the psychological evolution of early man has attracted a great deal of attention in recent years, as the importance of social behavior in animal life has come to be more and more appreciated. Several books summarizing the results of symposia and a number of journal articles which deal with this subject are listed at the end of this chapter. The discussion here will follow the detailed analysis given in this author's technical papers (Etkin, 1954, 1963*a*). It centers around the modifications in social behavior involved in the shift from the herbivorous ecology characteristic of old-world monkeys and apes to one in which hunting played a central role. Other points of view expounded in the literature are those of Oakley (1951, 1957, 1962) and Washburn (1960, 1962), who stressed the importance of tool-use and tool-making, and that of Chance (1953, 1962), who emphasized the socio-sexual relations. In accord with the viewpoint adopted in this chapter, the significance of social behavioral factors in the life of early man, particularly in relation to hunting, was noted by Bartholomew and Birdsell (1953), Eiseley (1956), and Imanishi (1961). Washburn and DeVore (1961) have also recognized the importance of hunting to social behavioral evolution in their most recent discussion.

The basic question is the nature of the ecological situation to which protocultural man was adapted. The *Pithecanthropus* fossil material and associated artifacts (Peking and Java man of about one-half million years ago) enable us to visualize fairly clearly the general ecology of that fossil man. He was a hunter of large and medium-sized animals, who used stone tools in the hunt and in other activities. Presumably he also made use of naturally occurring plant products for food. He used fire in the preparation of his food and took advantage of natural shelters, such as caves, for protection. Probably he fashioned rough clothes from animal skins. Thus with a cranial capacity of about nine hundred cubic centimeters, or roughly three-quarters of that of modern man, *Pithecanthropus* had definitely moved into the cultural sphere. In fact, as Weidenreich (1947) has pointed out, his mode of life was not too different from that of primitive hunting cultures of contemporary man.

The next, more lowly organized of fossil types is that of the South African ape men (*Australopithecus*), with a cranial capacity about half that of modern man (450–700 cc.) and not much above that of the larger apes. Our knowledge of its ecology is less clear. The South African paleontologists, particularly Raymond Dart (1960), who found and analyzed much of the fossil material favored the interpretation that this creature, too, lived in good part by hunting game animals, including baboons and antelope. Recent finds of paleoliths and bone industry associated with the *Australopithecines* by Dart (1960) and Leakey (Leakey and Des Bartlett, 1960) strongly support this view. Since these creatures were somewhat smaller than modern man and, like him, devoid of structural weapons (unlike apes, even their canines were not enlarged), hunting could only have been effectively carried out with pack organization and the use of weapons. From the structure of the pelvis, it can be said that the *Australopithecines* were bipedal and quite erect in locomotion. Their hands, therefore, were free for the use of weapons while they ran, in contrast to the situation in modern apes, which depend on their arms in rapid locomotion. We may conclude, therefore, that probably very early in evolution of his distinctive brain, man began the shift from a food-gathering to a pack-hunting ecology.

As we saw in considering the social organization of the wolf, the central condition for a pack-hunting mammal with young that have a considerable period of dependency is an integrated family organization, that is, one in which the male parent is an active co-operating member and one in which the maturing young remain associated with the parents. Such an organization contrasts strongly with that of the food-gathering macaque clan described above. Since man originated from such food-gatherers, in evolving in the direction of pack-hunting, protocultural man must have undergone a social behavioral revolution from something like the macaque type toward the wolf type. With this basic shift in mind, we can analyze the main trends in terms of the principal categories of social behavior used in this book.

### SOCIAL DOMINANCE

As in mammals generally, the persistence of the association of young with adults carries with it the continued dominance of the adults over the offspring, even when the latter are grown up—as we saw in the red-deer and wolf organizations. In contrast to the deer, however, the dominant male in hunters must be included as a co-operative member of the permanent social group. Furthermore, no exclusion of the young males from the group as occurs in the macaque could take place if a hunting pack is to be formed. Since the group is co-operative in hunting, dominance must tend toward the mild "leadership" type, such as in the female herd of the red deer, rather than the competitive aggressiveness of macaques. Competition in mating among males must, therefore, be avoided. This can be achieved by the suppression of mating in subordinates in the presence of dominants as we saw it to occur in wolves.

In this way the pack-hunting ecology would be expected to shift the mating system from the macaque type toward the integrated family with the elimination of mating competition among family members. Such a trend, developing in a large-brained primate already capable of considerable cultural regulation of social behavior, could give rise to the cultural incest taboos which are so fundamental a characteristic of human society and so unlike the condition of inbreeding in macaques. The essentials of such a system seem to obtain in wolves. In this view, the origin of incest taboos is to be sought in social behavioral factors rather than in such genetic considerations as the effects of inbreeding. It should also be noted that the recognition of biological paternity or other "blood" relationship is not a necessary condition for the establishment of incest taboos.

## TERRITORIALITY

There is a striking resemblance between territoriality as it appears in animal societies and the reactions of human groups to their "native" soil. One would not expect in hunting groups the extremely rigid defense of boundaries that characterizes some societies, such as song-bird pairs. Large predators seem to maintain a somewhat looser territoriality. The territory itself is necessarily large and often mobile, following the migration of game. The joining of small family groups to form a large pack, a well-authenticated phenomenon in wolves and thought also to occur in the gorilla (Kawai and Mizuhara, 1959; Schaller, 1963) indicates that a less rigid territoriality is to be expected in primitive man. Presumably, the groups that join have originated from a common family and have retained some effects of early socialization. A measure of such social acceptance between neighboring groups would furnish the basis for the shifting of individuals between groups. In this way exogamy, a necessary condition for the full operation of the incest taboo, might also be favored in the shift to hunting ecology.

### MATING BEHAVIOR

The absence of a definite estrus period is a distinctive feature of the female sex cycle in the human. Although we shall speak of a period of receptivity in the human female, it must be understood that this is not fully comparable to the receptivity of the estrus of infrahuman females, since the latter advertise their condition and actively seek mating. It has been pointed out by many writers that the constancy in sexual receptivity of the human female is one of the characteristics which make for permanent integration of the male into the family life. Our analysis of social behavioral factors enables us to specify ways in which the maintenance of this constant state of receptivity upon the part of the female favors the integration of the male into the family organization. It will be recalled that in mammals generally, sexual congress is limited to the estrus period of the female by her behavioral reactions, whereas the male maintains constant sex interest. The human female is continuously sexually receptive. The sustained sex interest of the partners thus favors the integration of the male (Zuckerman, 1932; Chance, 1962). Furthermore, in the absence of a definite estrus period, the specific stimuli furnished by the usual mammalian female as a sign of estrus (i.e., odors, sexual skin color, etc.) are no longer functional and have dropped out. This must minimize the occasions for male rivalry for the females and favor a generalized socialization process as the primary determinant of sexual activity. As we have seen, play is such a general socializing activity. Sexual play in humans is greatly facilitated by the assumption of the upright posture, which makes ventral copulation possible, replacing the awkward mounting posture found in other primates (Fig. 10.13).

The importance of socialization between mates as a determinant of sexual activity in man in contrast to the direct stimulation of sexual activity by specialized signaling devices at estrus gives further emphasis to the point made above—that sexual activity tends to be governed by dominance and other purely social relations. This, of course, furnishes the basis for the establishment of the cultural taboos that are so prominent a part of human sexuality. The detailed studies on Japanese monkeys (Itani, 1958; Kawai, 1958) reveal a high degree of complexity of social behavior in which consort relations interact closely with the hierarchical structure of the group. Each male, for

FIG. 10.13. These sketches illustrate the positions assumed by the gorillas in copulation. (Drawings by G. Schaller, *The Mountain Gorilla,* 1963.)

example, tends to have stable consort relations, and the social position of the consort and even of her offspring is a function of the place of the male in the hierarchy, thus adumbrating social class structure. It can readily be appreciated that an evolutionary pressure toward the integrated family would greatly expand such relations and help make the transition to a culturally determined pattern of socio-sexual activity, as we find it in human societies.

### PARENTAL CARE AND DIMORPHISM

The principal basis for the evolution of dimorphism in animal societies, as we have seen, is the difference in reproductive roles of the two sexes. This is well illustrated in macaque society by the development of a high degree of aggressive potential, powerful canine teeth, large size and strength, and aggressive disposition in the males, in contrast to the females. Some students have attempted to see in man a behavior structure derivative of this, on the assumption of some sort of evolutionary carry-over from the more general primate conditions. If the general thesis of this book is correct and behavior systems are readily modifiable in evolution by ecological demands, then we should look more to the wolf than to the macaque as a basic model for understanding protocultural man. We found that the male wolf was integrated into the family and provided parental care in correlation with the necessities of the pack-hunting situation, particularly the dependence of the pregnant and nursing female upon other members of the pack. In the human situation, as visualized in this theory, the dependence of the female is further emphasized by the greater size and helplessness of the human infant, factors which would preclude any hunting on the part of the female for most of her active breeding life. This, of course, correlates with the economic role of the female in human hunting societies as food-gatherer and domestic. In marked contrast to the macaque situation, a strong selection pressure must be presumed to have operated toward reducing the aggressiveness of the males within the family situation.

In large predaceous animals group size is generally quite small in contrast to the large groupings of foraging animals such as monkeys. Small group size also reduces the competition between males for females and so limits the trend toward dimorphism in early man. Since only the males could hunt, they must have evolved in the direction of co-operation with all members of the group. They must co-operate with other males in hunting; they must bring back food for the women and children and assume other responsibilities, such as training the boys in the techniques of hunting.

The difference in economic roles visualized for males and females would tend to maintain some differences between the sexes in respect to body build, strength, and endurance, but without encouraging intragroup competition among males. The absence of enlargement in the canines of australopithecine males strongly suggests the absence of competition between males for females. Thus the dimorphism which characterizes modern and ancient men and ape men, namely, moderate differentiation in strength but without strongly developed aggressive potential, is here visualized as adaptive in the ecology of protocultural man. It fits the division of labor between the sexes better than would a sexually competitive pattern among males such as exists in macaque society.

### PROLONGATION OF DEPENDENCY

The third distinctively human characteristic in reproduction to be accounted for is the extreme lengthening of the period of growth and learning. In primitive societies the human infant nurses until at least two to three years of age; the monkey baby, for only two or three months; and the chimpanzee, for less than a year. Puberty is correspondingly delayed. As has been repeatedly pointed out, the prolongation of the period of dependency is an essential adaptation of the human condition, making for the educability of the child and therefore for human culture. But what we must also realize is that this could only arise as a continuing evolutionary process *pari passu* with the development of the integrated family ecology; for, unless the male "breadwinner" were incorporated into the family, the burden on the female of having several dependent young would be unworkable. It should be recognized that in an ecology such as that of the macaque, where the young are a burden upon the female only and where she must forage for herself and baby, there is a strong selection pressure for rapid development of the young, just as there is in birds generally (see Chapter 4). The condition of slower development in man may therefore be viewed as the result, in part, of a relaxation of the extreme pressure for rapid development as well as a positive pressure for a longer learning period. This means, then, that the prolongation of the human period of immaturity may have started independently of its value in a cultural system, but, once started, it became accelerated by the fact that within the integrated family situation it permitted the further advantage of cultural transmission.

The final category of social behavior to be analyzed is that of socialization; this has already been touched upon above. Of the factors which may be subsumed under this heading, we should consider three at this point: language, play, and cultural transmission, as presenting distinctively human problems.

### LANGUAGE

The central aspect of the problem of language, from the biological viewpoint may be stated as follows: What selective forces led to the development of the use of symbols in communication in man, whereas such use has never evolved appreciably in even the highest apes? It is commonplace in anthropological discussion to point to the enlargement of the human brain as making speech possible. Generally the basis of the enlargement of the brain is ascribed to other selective forces—tool-use, control of autonomic centers, etc. The selectionist viewpoint espoused here, however, holds that complex structures like complex behavior represent responses to selection pressures specifically and directly favoring them. They do not evolve as incidental accompaniments of other selection pressures. If tool-use, for example, were valuable for foraging animals (which is doubtful, since foraging primates make almost no use of it in food-getting, although quite capable of using sticks, etc., in defense), the kind of mentality it would favor would make for tool-use, not for language (Figs. 10.14 and 10.15).

The concept of an integrated family organization, with economic differentiation of male and female roles as here suggested for proto-

**FIG. 10.14.** The Galapagos woodpecker finch is using a stick to dislodge insects from a crevice. Though this must be considered a type of "tool-use," presumably it is an inherited rather than culturally determined behavior and is inflexible. (After D. Lack, *Darwin's Finches,* 1947.)

FIG. 10.15. Chimpanzees have a strong tendency (apparently innately determined) to manipulate objects of appropriate sizes by turning them over, stuffing them into available cavities, etc. This is done most effectively in play rather than when the animal is attempting to reach a goal in problem solving experiments. Instances which appear to be "insightful" solutions of problems such as that of fitting two sticks together to reach an object out of reach of either may represent mere coincidences resulting from this manipulative tendency. (After P. Schiller, *in* C. H. Schiller (ed.), *Instinctive Behavior,* 1957.)

cultural man, points to an important difference from the condition of other primates, which may account for the selection pressure for symbol-formation in man and not in them. In the monkey or ape, each individual forages for himself—except, of course, for infants. For no important period of time are the members of the group out of one another's range in terms of communication. Hence, signals rather than symbols suffice for communication (Etkin, 1963*b*). For example, should an enemy be detected by one monkey, its outcries would serve to arouse the attention of other members who, being in the immediate vicinity, could react appropriately. It is not necessary for the first animal to convey information of a specific kind about objects not present or to specify the direction or nature of the source of alarm. It is sufficient for the animal to express his own alarm and flee in his own way. Thus a signal expressing the sender's own condition, rather than a symbol with a specific referent, suffices. As we have seen in Chapters 7 and 8, such signals are widespread among all social animals.

In the integrated family organization the division of labor necessarily keeps the members out of touch. Communication about the hunt—for example, the slaying of a large animal at a distance, requiring co-operation for handling or the injury of a member of the pack—must be transmitted from the hunters to the others. Only symbols conveying specific information about referents not otherwise to be perceived by the recipient will do this. Conversely, events on the domestic scene must be conveyed to the returned hunters. In short, the integrated family with two separate areas of action depends upon the ability to communicate information about things not immediately to be perceived. We may suppose that the special conditions of the integrated-family ecology gave rise to the selection pressure whose outcome was the use of sounds as symbols with specific referents. This communication system arose in one species only because the selection pressure operated upon an animal which was already highly intelligent and which adopted a new mode of living. This unique combination of factors occurred only in the evolution of man.

### PLAY

Something has already been said about the special aspects of human play in connection with reproduction. In man as in wolves, the adults, both male and female, play with each other and with the young. Play in man is not only physical but verbal. Social conversation is one of the most important ways of maintaining a friendly social bond and the basis of familiarization (Etkin, 1963*b*).

Another characteristic of human play is its relation to training for adult activities. The familiar fact that the play of children so often is an imitation of adult activities must be viewed as part of the adaptation of man to cultural transmission. This spontaneous play grades over into definitive instruction and practice of adult roles in which the differentiation of male and female is prominent. As we have previously emphasized, the drives and capacities for learning in particular ways are part of the adaptive mechanisms of an animal. The prominence of the drives to assume and practice adult roles is thus part of man's adaptation to his cultural mode of life. It differs, of course, only in degree from that seen in many mammals. The hunting play of many predaceous mammals, which also includes definitive instruction by the parent, is a case in point. Another is the observation of Fraser Darling that the female yearling in the red deer attend to the reactions of their mothers toward weather signs, whereas the males, never destined to be herd leaders, do not attend and apparently never learn to be proficient.

### CULTURAL TRANSMISSION

Cultural transmission, the predominant characteristic of human mentality, has a firm if narrow base in animal behavior. As we have seen, even among birds there are clear examples of the transmission of learned behavior from one member of a group to another (cultural transmission). The learning of foodstuffs and enemies is widespread. One of the most curious examples of such learning is that of the tits and various other British birds which have learned to rob milk bottles by stripping or breaking through the paper tops. This technique appears to have been started by a few birds in an area and spread to the others in the manner typical of the diffusion of cultural elements. As in other cases discussed in Chapter 7, this activity is merely a modification of the normal behavior pattern of these animals, which find part of their food by stripping bark. What has been learned, therefore, is the new stimulus situation releasing the old motor response (Fisher and Hinde, 1949).

In point of fact, all learning involving social relations necessarily assumes somewhat the character of cultural transmission. For example, when a dove is imprinted to its own species and subsequently mates "correctly," it thereby sets up the situation for transmitting this same behavior to its offspring. Whether we choose to call this cultural or "pseudo-cultural" transmission, we must recognize that it has much the same effect as direct teaching in that one generation transmits its learned patterns to the next. Such "pseudo-cultural" transmission appears to play a considerable role in territory, mating, and perhaps other activities in birds.

In mammals cultural transmission of a more conventional character is more commonplace. Since the large social mammals are long-lived, there is much occasion for the direct learning of stimuli regarding territory, migratory paths, food, and enemies. Mammalian action patterns, being less rigidly organized, also are more subject to cultural modification, as we saw in discussing learned hunting techniques. Primates show cultural transmission most extensively. This was particularly evident in the reports from the Japanese Monkey Center [Fig. 10.16]. In apes, though little is known of the details of their behavior in the wild, such transmission of learned behavior is very apparent in laboratory and zoo conditions. Apparently, chimpanzees cannot even mate successfully without instruction or guidance from experienced animals (see Chapter 5, page 139).

If, then, we think of protocultural man under the circumstances (presumably of climatic change) that were forcing him toward a hunting ecology, we may recognize that an alternative to the usual biological method of evolution presented itself to him. Instead of

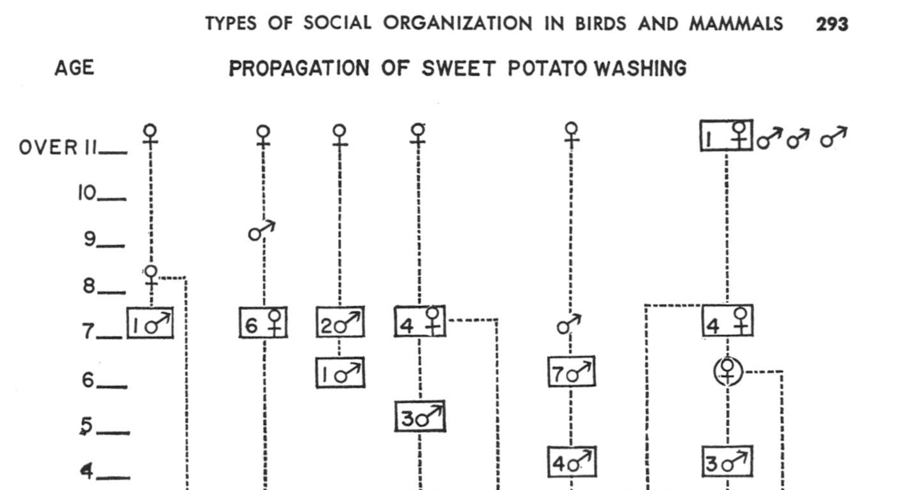


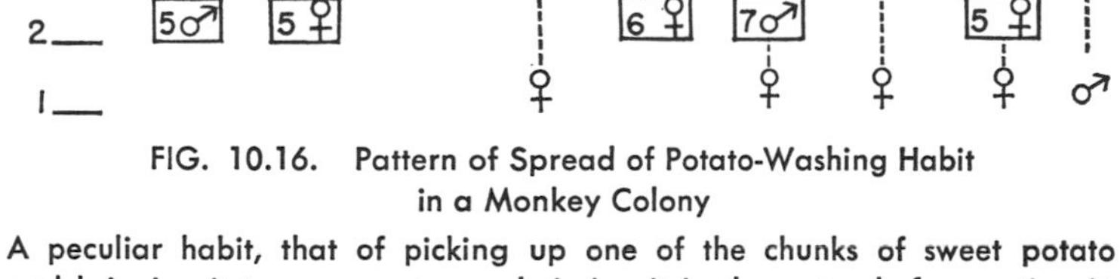

FIG. 10.16. Pattern of Spread of Potato-Washing Habit in a Monkey Colony

A peculiar habit, that of picking up one of the chunks of sweet potato and bringing it to some water and rinsing it in the water before eating it, was started in one of the colonies of Japanese monkeys by a young female (*encircled*). This habit was seen to be picked up first by her mother and two playmates of comparable age. It then spread to other young animals of the colony, primarily among playmates, but at the time of writing had not spread to any of the older females or males. The lines indicate genealogical relations from females. The animals taking up the habit are inclosed in boxes and the sequence in which the habit was assumed is given by the inclosed numbers. (After Miyadi, *Proc. XV Int. Cong. Zool.*, 1958.)

altering each of the behavioral characteristics individually by genetic change, a good part of the behavioral transition could be achieved by further development of his already extensive capacity for cultural transmission of learning. Let us consider dominance behavior as an example. Instead of altering this behavior mechanism in respect to the stimuli eliciting it and the motor patterns of its expression by genetic change of neural mechanisms, the general dominance drive could be made subject to learned stimuli and methods of expression. This would lead toward the situation as we see it in man, where both the objects and the manner of expression of dominance drives vary with each culture. In this way the alteration of behavioral patterns by cultural means would become the basis of human evolution. Such a process of change would be vastly more rapid than the usual genetic method. Furthermore, it would be more plastic, permitting adjustment to different ecological niches by cultural variations of the same basic drives. Such an evolutionary pattern, as it developed, would demand greatly expanded neural resources for the generalized learning involved. Genetic change leading to increased size of the brain would thus be favored.

At the australopithecine level we might expect that the behavioral shift was still in good part dependent on genetic change and would involve only slight increase in neural complexity or brain size. Merely to shift the pattern of male aggressiveness to the integrated type requires in itself no increase in brain size, as seen from the fact that wolves show this characteristic. But as cultural transmission becomes more important and culturally determined behavior accumulates, a feed-back system magnifies the difference between those groups in which it takes hold and those still dependent upon conventional evolutionary mechanisms. This then becomes an autocatalytic system continually accelerating the selection pressure for further shift to the cultural mode of evolution. It is this dependence upon cultural regulation of all drives that requires increase in the cerebral cortex. Such an autocatalytic system leads to the explosive expansion of the brain that appears to have taken place within a few hundred thousand years early in the Pleistocene period (Eiseley, 1956). The outcome is a modern man who appears to have the same drives as we noted in other animals but little in the way of innate mechanism to recognize the appropriate stimuli or to perform the motor patterns for their expression. What he does have is an enormous predisposition to learn these things from others and to teach them to another generation. Thus the nature of man is to appear to have no fixed nature, only tradition.

Of course, the above remarks will be recognized as highly speculative. They have been included merely to bring out the possibilities of interpretation which consideration of social-behavioral factors add to our resources for comprehending human behavior and evolution. It is this author's opinion that an understanding of human nature must be founded upon an understanding of social behavior in lower animals. Progress in this direction depends, not upon reliance on occasional analogies, but upon recognizing the vast variety of forms that the same fundamental principles can take in the kaleidoscope of nature. It is to be hoped that an appreciation of the ecological principles and the physiological mechanisms in the social behavior of animals will enable us to see more clearly the possibilities for the analysis of our own behavior.

## REFERENCES

### BIRDS

ARMSTRONG, E. 1947. *Bird Display and Behavior*. New York: Oxford University Press.

HEINROTH, O., and HEINROTH, K. 1958. *The Birds*. Ann Arbor: University of Michigan Press.

HOWARD, E. 1948. *Territory in Bird Life*. 2d ed.; London: William Collins Sons & Co.

KENDEIGH, S. C. 1952. *Parental Care and Its Evolution in Birds*. Urbana, Ill.: University of Illinois Press.

LACK, D. 1943. *The Life of the Robin*. London: H. F. & G. Witherly. Reprinted, 1953; London: Penguin Books.

NICE, M. M. 1937, 1943. Studies in the life history of the song sparrow, I, II. *Trans. Linnaean Soc.* (N.Y), **4:** 1–246; **6:** 1–239.

TINBERGEN, N. 1939. Field observations of east Greenland birds. *Trans. Linnaean Soc.* (N.Y), **5:** 1–91.

———. 1953. *The Herring Gull's World*. London: William Collins Sons & Co.

WHITMAN, C. O. 1919. *The Behavior of Pigeons,* ed. H. CARR. Publ. No. 257, pp. 1–161. Washington, D.C.: Carnegie Institute.

### MAMMALS

BOLWIG, N. 1959. A study of the behavior of the Chacma baboon, *Papio wisinus. Behaviour,* **14:** 136–63.

BOURLIERE, F. 1954. *The Natural History of Mammals*. New York: Alfred A. Knopf.

CALHOUN, J. B. 1948. Mortality and movement of brown rats (*Rattus norvegicus*) in artificially supersaturated populations. *J. Wildlife Monogr.,* **13:** 167–72.

———. 1952. The social aspects of population dynamics. *J. Mammal.,* **33:** 139–59.

CARPENTER, C. R. 1934. A field study of the behavior and social relations of the howling monkeys. *Comp. Psychol. Monogr.,* **10:** No. 2.

———. 1942. Sexual behavior of free ranging rhesus monkeys (*Macaca mulatta*). *J. Comp. Psychol.,* **33:** 133–62.

DARLING, F. F. 1937. *A Herd of Red Deer*. London: Oxford University Press.

DAVIS, D., EMLEN, J., and STOKES, A. 1948. Studies on home range in the brown rat. *J. Mammal.,* **29:** 207–25.

IMANISHI, K. 1957. Social behavior in Japanese monkeys, *Macaca fuscata. Psychologia,* **1:** 47–54.

MURIE, A. 1944. *The Wolves of Mount McKinley* (Fauna of the National

Parks of the U.S., No. 5). Washington, D.C.: U.S. Government Printing Office.

SCHALLER, G. 1963. *The Mountain Gorilla.* Chicago: University of Chicago Press.

ZUCKERMAN, S. 1932. *The Social Life of Monkeys and Apes.* New York: Harcourt, Brace.

HUMAN EVOLUTION

*Symposia*

GAVAN, J. (ed.). 1955. *The Non-Human Primates and Human Evolution.* A symposium. Detroit: Wayne University Press. (See particularly the articles by CARPENTER, NISSEN, HAYES and HAYES.)

MONTAGU, M. (ed.). 1962. *Culture and the Evolution of Man.* A symposium largely of reprinted articles. New York: Oxford University Press. (See particularly the articles by OAKLEY, WASHBURN, BARTHOLOMEW and BIRDSELL, CHANCE, ETKIN, DOBZHANSKY and MONTAGU, METTLER, HALLOWELL, EISELEY, and MONTAGU.)

ROE, A., and SIMPSON, G. G. (eds.). 1958. *Behavior and Evolution.* A symposium. Detroit: Wayne State University Press. (See particularly the articles by COLBERT, PRIBRAM, NISSEN, HARLOW, THOMPSON, WASHBURN, and AVIS, HUXLEY, FREEDMAN and ROE, and MEAD.)

SPUHLER, J. N. (ed.). 1958. *Natural Selection in Man.* A symposium. Detroit: Wayne State University Press. (See particularly the article by NEEL.)

———. 1959. *The Evolution of Man's Capacity for Culture.* A symposium. Detroit: Wayne State University Press. (See particularly the articles by SPUHLER, WASHBURN, and SAHLINS.)

TAX, SOL (ed.). 1960. *The Evolution of Man.* A symposium. Chicago: University of Chicago Press. (See particularly the articles by LEAKEY, WASHBURN and HOWELL, CRITCHLEY, and HALLOWELL.)

WASHBURN, S. (ed.). 1961. *Social Life of Early Man.* A symposium. Chicago: Aldine Publishing Co. (See particularly the articles by BOURLIERE, CHANCE, HEDIGER, SCHULTZ, WASHBURN, and DE VORE, OAKLEY, HALLOWELL and CASPARI.)

*Journal Articles*

BARTHOLOMEW, G., and BIRDSELL, S. 1953. Ecology and the protohominids. *American Anthropologist,* **55**: 481–98. Reprinted in A. MONTAGU (ed.), 1962. (See symposia listed above.)

CHANCE, M., and MEAD, A. 1953, 1962. Social behavior and primate evolution. *Symposia of Society for Experimental Biology,* **7**: 395–439. A revision of this article by Chance is included in A. MONTAGU (ed.), 1962. (See symposia listed above.)

DART, R. 1960. The bone tool-manufacturing ability of *Australopithecus prometheus. American Anthropologist,* **62**: 134–43.

EISELEY, L. 1956. Fossil man and human evolution. *In:* THOMAS (ed.),

*Yearbook of Anthropology,* 1955. Reprinted in A. MONTAGU (ed.), 1962. (See symposia listed above.)

ETKIN, W. 1954. Social behavior and the evolution of man's mental faculties. *Amer. Nat.,* **88:** 129–42. Reprinted with additional discussion in A. MONTAGU (ed.), 1962. (See symposia listed above.)

———. 1963*a*. Social behavioral factors in the emergence of man. *Human Biology,* **35:** 299–311.

———. 1963*b*. Animal communication. *In:* J. EISENSON (ed.), *Psychology of Communication.* New York: Appleton-Century Crofts.

FISHER, J., and HINDE, R. A. 1949. The opening of milk bottles by birds. *Brit. Birds,* **42:** 347–57.

IMANISHI, K. 1961. The origin of human family (English summary). *Jap. J. Ethnology,* **25:** 119–38.

ITANI, J. 1958. On the acquisition and propagation of a new food habit in the natural group of the Japanese monkey at Takasaki Yama. *J. Primatology,* **1:** 84–99.

KAWAI, M. 1958. On the rank system in a natural group of Japanese monkey, I. *J. Primatology,* **1:** 111–31.

KAWAI, M., and MIZUHARA, H. 1959. An ecological study on the wild mountain gorilla (*Gorilla gorilla beringei*). *J. Primatology,* **2:** 1–43.

LEAKEY, L., and DES BARTLETT. 1960. Finding the world's earliest man. *Nat. Geog.,* **118:** 420–35.

OAKLEY, K. 1957. Tools makyth man. *Antiquity,* **31:** 199.

———. 1951, 1962. A definition of man. *Science News,* 20. Reprinted in A. MONTAGU (ed.), 1962. (See symposia listed above.)

YERKES, R. M. 1943. *Chimpanzees: A Laboratory Colony.* New Haven, Conn.: Yale University Press.

WASHBURN, S. 1960, 1962. Tools and human evolution. *Sci. Amer.,* **203:** 63–75. Reprinted in A. MONTAGU (ed.), 1962. (See above and also symposia article in A. ROE and G. G. SIMPSON [eds.], 1958.)

WASHBURN, S., and DE VORE, I. 1961. Social behavior of baboons and early man. *In:* S. WASHBURN (ed.), *Social Life of Early Man.* Chicago: Aldine Pub. Co.

WEIDENREICH, F. 1947. The trend of human evolution. *Evolution,* **1:** 221–36.

# Index